The Math Chat Book

$1000 Math Chat Book Quest

To solve this Quest, thy self must look
To Problems deep and in the Book;
One Lead doth tell thee what to Give for what be Broken,
Nor Light, nor Spoil, nor Speech, nor Token.

Send your one-paragraph explanation story to Frank.Morgan @williams.edu with bibliographic citation. Best correct response which stands for a year (or maybe earlier) wins $1000 award. No particular mathematical training is required. You can check on the current status of this contest at Math Chat at www.maa.org.

Frank Morgan

Cover illustration: Views of a fantastic new computer-simulated double soap bubble, which has threatened to overturn mathematicians' assumptions (see Episode 23). Graphics by John M. Sullivan, University of Illinois. Cover design by Freedom by Design.

Library of Congress Catalog Card Number 99-067970

ISBN 0-88385-530-5

Printed in the United States of America

Current printing (last digit):
10 9 8 7 6 5 4 3 2 1

The Math Chat Book

Frank Morgan

Illustrated by James F. Bredt

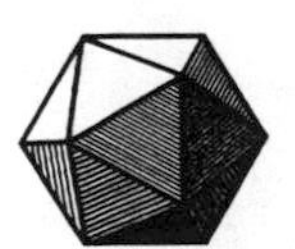

Published and Distributed by
THE MATHEMATICAL ASSOCIATION OF AMERICA

SPECTRUM SERIES

Published by
THE MATHEMATICAL ASSOCIATION OF AMERICA

SPECTRUM SERIES

The Spectrum Series of the Mathematical Association of America was so named to reflect its purpose: to publish a broad range of books including biographies, accessible expositions of old or new mathematical ideas, reprints and revisions of excellent out-of-print books, popular works, and other monographs of high interest that will appeal to a broad range of readers, including students and teachers of mathematics, mathematical amateurs, and researchers.

All the Math That's Fit to Print, by Keith Devlin
Circles: A Mathematical View, by Dan Pedoe
Complex Numbers and Geometry, by Liang-shin Hahn
Cryptology, by Albrecht Beutelspacher
Five Hundred Mathematical Challenges, Edward J. Barbeau, Murray S. Klamkin, and William O. J. Moser
From Zero to Infinity, by Constance Reid
I Want to be a Mathematician, by Paul R. Halmos
Journey into Geometries, by Marta Sved
JULIA: a life in mathematics, by Constance Reid
The Lighter Side of Mathematics: Proceedings of the Eugène Strens Memorial Conference on Recreational Mathematics & its History, edited by Richard K. Guy and Robert E. Woodrow
Lure of the Integers, by Joe Roberts
Magic Tricks, Card Shuffling, and Dynamic Computer Memories: The Mathematics of the Perfect Shuffle, by S. Brent Morris
The Math Chat Book, by Frank Morgan
Mathematical Carnival, by Martin Gardner
Mathematical Circus, by Martin Gardner
Mathematical Cranks, by Underwood Dudley
Mathematical Fallacies, Flaws, and Flimflam, by Edward J. Barbeau
Mathematical Magic Show, by Martin Gardner
Mathematics: Queen and Servant of Science, by E. T. Bell
Memorabilia Mathematica, by Robert Edouard Moritz
New Mathematical Diversions, by Martin Gardner
Non-Euclidean Geometry, by H. S. M. Coxeter
Numerical Methods that Work, by Forman Acton
Numerology or What Pythagoras Wrought, by Underwood Dudley
Out of the Mouths of Mathematicians, by Rosemary Schmalz

Penrose Tiles to Trapdoor Ciphers . . . and the Return of Dr. Matrix, by Martin Gardner

Polyominoes, by George Martin

The Random Walks of George Pólya, by Gerald L. Alexanderson

The Search for E. T. Bell, also known as John Taine, by Constance Reid

Shaping Space, edited by Marjorie Senechal and George Fleck

Student Research Projects in Calculus, by Marcus Cohen, Arthur Knoebel, Edward D. Gaughan, Douglas S. Kurtz, and David Pengelley

The Trisectors, by Underwood Dudley

Twenty Years Before the Blackboard, by Michael Stueben with Diane Sandford

The Words of Mathematics, by Steven Schwartzman

To order MAA publications, contact:

MAA Service Center
P. O. Box 91112
Washington, DC 20090-1112
800-331-1622 FAX 301-206-9789

Contents

Introduction

"Math Chat" is a live call-in TV show, a column (originally in *The Christian Science Monitor* and now at the Mathematical Association of America web site www.maa.org), and now a book. Math should be interesting and fun, so we have questions, answers, and prizes. Indeed, I'll be awarding prizes for the best comments and new questions I get from you readers of this book.

Many Americans feel excluded from mathematics. I attended a dinner with some college students. One woman complained bitterly about the terrible experience she had had with mathematics in high school. A man told a similar tale of woe. The rest could hardly wait to tell their own miserable stories. I did not know what to say. Then another student told about a math teacher's kindness. Everyone was quiet. The first woman started to cry. She felt that she had been cheated out of something wonderful. Everyone likes math, when they think they have a chance to participate. Our nation deserves the chance.

Everyone deserves to enjoy math. A question about a teenager who is good at everything but math appeared in an "Ask Marilyn" column (*Parade*, Febuary 15, 1998). Marilyn recommended concentrating on the student's strengths. We agree, but we would also encourage finding something related to math the student could enjoy, whether or not it has anything to do with schoolwork. Everyone deserves to enjoy something of mathematics.

People do like math when they think they can understand it. Kids, housewives, retired folks stop me on the street or in the supermarket to tell me how much they enjoy "Math Chat" on TV, even though they parted with math long ago or never liked math. Unfortunately, most have had experiences with math that once made them feel stupid or inadequate. They don't realize that mathematicians feel exactly the same way, at least until they find a way to laugh off the endless barrage of mistakes and fears and just have some fun.

Math Chatters include school children, college students, retired scientists, lawyers, and prison inmates. When I sent a prison inmate an award for a published contribution, the warden wrote back:

> Dear Professor Morgan:
>
> The Department of Corrections prohibits inmates from entering contests and winning prizes. As such, we are returning the prize you sent.

Oh well.

The following old puzzle was submitted by Thomas Linton, who is scripting such stories for a cartoon book, and also by Joe Herman.

Hotel Paradox. Three guys go into a hotel, each with $10 in his pocket. They book one room at $30 a night. A short while later a fax from headquarters directs the hotel to charge $25 a night. So the receptionist gives the bellhop $5 to take to the three guys sharing the room. Since the bellhop never got a tip from them and because he can't split $5 three ways, he decides to pocket $2 and give them each one dollar back (Figure 1). So each of the three guys has now spent $9 and the bellhop has $2, for a total of $29. Where's the extra dollar?

Answer. The bellboy's $2 was included in the $27 the guests paid, so it should not be added but *subtracted*, yielding the remaining $25 the hotel

FIGURE 1 The famous hotel paradox. Where is the extra dollar?

got. James Turner notes that "adding $2 to $27 is meaningless, and only seems reasonable because the 'answer' of $29 is so close to the original number, $30." Joe Shipman concludes that "this is a good puzzle because the faulty step is so brazenly stupid it can be hard to notice, sort of like Poe's purloined letter which was hidden in plain sight on a table." Robin Konicek suggests that "Tatum O'Neal and her father Ryan could well have used this in the movie *Paper Moon* to try to swindle some more folks."

Readers find ways to tell us gently that this is an old puzzle. Aubrey Dunne writes, "My father had posed this problem to me prior to WWII."

The first "Math Chat" prize is the classic book *Flatland* by Edwin A. Abbott, which helps you understand four-dimensional space by thinking about how to explain our own three-dimensional space to two-dimensional creatures. Not everyone likes the book. Tarah Smith Nellis writes:

> I borrowed a copy from a library and was initially delighted by the author's cleverness. However, about 20 or so pages into the novel, I was dismayed to encounter a severe put down of the intelligence of women.

Actually, as the second of our award books, *The Mathematical Tourist* by Ivars Peterson, explains, "*Flatland* is a sharply delineated satire, which reflected widely debated social issues in Victorian Britain. Abbott was a strong advocate of women's rights, and he couldn't resist taking a satirical swipe at his society's attitudes toward women." Indeed, one edition of *Flatland* announces on the front cover: "Humour, satire, logic, all combined in a brilliantly entertaining classic of the fourth dimension"; and on the back cover: "In a world where women resemble needles ... and men are various polygons who recognize each other by their angles (from the wedge-like soldiers to the perfect circle), social satire of Victorian mores is hilariously sharp and potent."

I like to use *Flatland* as a lead-in to another award book, *The Boy Who Reversed Himself* by William Slater, a modern children's book which gives a fascinating speculative account of all higher dimensions, and in which the starring character is a girl. Mathematics is for everyone.

The movie *Good Will Hunting* tells about a prize-winning MIT mathematician and a young prodigy, played by Matt Damon. The credits list Harvard physicist Sheldon Glashow and MIT mathematician Daniel Kleitman. According to Kleitman:

> The original story had the hero a brilliant physicist; they contacted Glashow, who said that it didn't ring true as a physicist; he suggested they make him a mathematician and suggested they talk to me. They told me they wanted to hear mathematicians talk .... I muttered a few words, got ahold of a post doc [recent Ph.D.], Tom Bohman, and had a discussion with him of various results and topics and things. They busily took notes; they seemed quite alert and intelligent, though mathematically illiterate. After a while (an hour? two hours? who can remember) they left, thanking us profusely for our help.

Kleitman appears briefly in the movie, outside the window of the Tasty while the stars are talking at the counter.

"Math Chat" began with the TV show on Willinet cable in Williamstown, Massachusetts, produced with Williams College students Eric Watson, Aaron Dupuis, Charmaine Mangram, and Liz Claflin. The Princeton University staff included Derek Smith, Ralph Thomas, Mike Lindahl, Ari Turner, Jade Vinson, Adrian Banner, and Ray Marquette. I also thank my "Math Chat" newspaper column editors at *The Christian Science Monitor* Eric Evarts, Kim Campbell, and Susan Leach, and my editor at the Mathematical Association of America, Carol Baxter.

Enjoy math. It's for everybody.

Frank Morgan

Allentown, Pennsylvania
September 17, 1998
Frank.Morgan@williams.edu

The TV and column versions of "Math Chat" are both available via Prof. Morgan's homepage at http://www.williams.edu/Mathematics. We especially recommend the March 9, 1998 TV show, which features high school winners of his Soap Bubble Geometry Contest.

(a)

(b)

FIGURE 2 (a) The original Williams College student "Match Chat" TV staff (left to right): Ted Melnick '99, Eric Watson '97, Liz Claflin '99, the author and host, Charmaine Mangram '99, and Aaron Dupuis '99. (b) Derek Smith *99 led the "Math Chat" staff at Princeton.

FIGURE 3 The "Math Chat" TV show, Princeton University, March 9, 1998. Co-host Mykel Kulkarni, Stacy Pierre and Shanay Gillette (winners of Morgan's Soap Bubble Geometry Contest, Riverside Elementary School), and host Frank Morgan.

First Story: Time

EPISODE *1*

Riddle O'Clock

The following riddle provokes magnificent struggles by readers.

Riddle O'Clock. John lives in an Atlantic Coast state of the United States, and Mary lives in a Pacific Coast state. When talking on the telephone from home, they realized it was the same time in both locations. How could this have been possible?

Since Eastern time and Pacific time differ by three hours, this puzzle sounds triply difficult. Many readers try to use Alaska or Hawaii somehow, but they are just an hour or two earlier than California, just four or five hours earlier than New York, nowhere near the 12 hours earlier needed to carry them back around the clock to resemble Eastern time. In desperation some readers call Texas an Atlantic coast state or argue for foreign Pacific Coast states such as Mexico, Panama, Columbia, Equador, Peru, or Vietnam. Rest assured that "Math Chat" does not deal in such trickery.

The first key to the puzzle is that the panhandle of Florida, which is certainly an Atlantic Coast state, extends well into the Central time zone. John probably lives in Pensacola, Florida. That's one hour.

Secondly, one Pacific Coast state extends into the Mountain time zone: not California, not Washington, but Oregon, perhaps to share time with neighbors in Idaho. Mary probably lives in Ontario, Oregon. That's two hours, and John and Mary's clocks are now just one hour apart, as in Figure 4.

The key to the third hour is that daylight saving time ends an hour earlier in the Central time zone than in the Mountain time zone. On the last Sunday in October, for the hour following the turn-back from 2 a.m. Central Daylight Time back to 1 a.m. Central Standard Time (CST), which agrees with 1 a.m. Mountain Daylight Time (MDT), John and Mary would enjoy the same time in both locations.

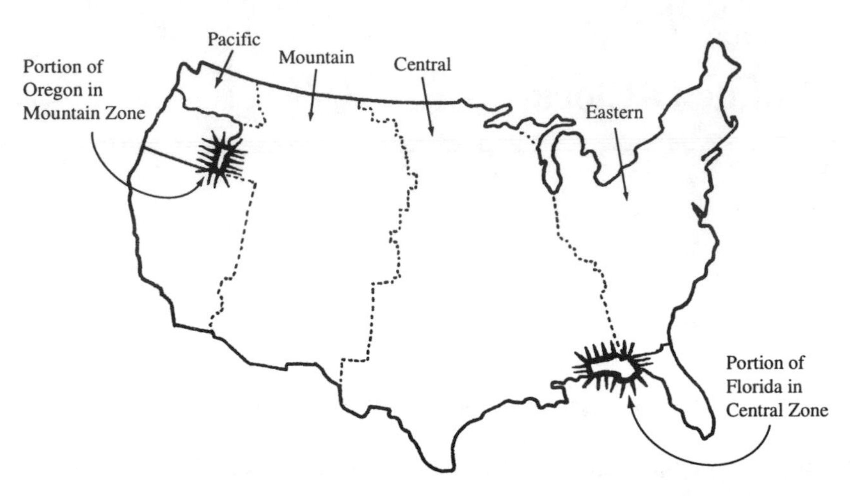

FIGURE 4 Florida and Oregon, although on opposite coasts, have portions in neighboring Central and Mountain time zones. That still leaves one hour's discrepancy to resolve.

An amazing feature of this solution is that it tells us almost exactly where and when this remarkable coincidence occurred: Pensacola, Florida and Ontario, Oregon, on the last Sunday in October, at about 1:30 a.m. (CST = MDT).

EPISODE 2

Does the Sun Rise in the East?

Everyone has heard that the sun rises in the east, but does it rise due east? A bit south of east? A bit north of east? Does it depend on where you live? Does it depend on the time of year?

I always thought that in the northern hemisphere, at least north of the tropic of Cancer (about 23 degrees north latitude), the sun always rose south of east. The Tropic of Cancer is the northern limit of the sun's apparent journey, reached around June 21, and the Tropic of Capricorn (about 23 degrees south latitude) is the southern limit of the sun's journey, reached around December 21.

Well my parents, my brother David, and I were at the Jersey shore one summer, and my mother claimed that the sun was rising *north* of east. I explained to her why that was impossible and eagerly awaited the next morning to prove my point. To my amazement the sun rose in the northeast! The fact of the matter is that when it is spring and summer in the Northern Hemisphere, when the North Pole is tilted toward the sun as in Figure 5, the sun rises north of east at every location on earth (except around the North Pole, where it never sets, and around the South Pole, where it never rises). Similarly, when it is fall and winter in the Northern Hemisphere, when the South Pole is tilted toward the sun, the sun rises south of east at every location. On the first day of spring or fall, there is no tilt, and the sun rises due east everywhere.

My original mistake was worrying too much about the point on the earth directly under the sun. All that really matters is the direction to the sun from the earth, largely determined by the earth's tilt. In my defense, however, it should be noted that if you live above the Tropic of Cancer, at high noon the sun is always to the south, at all times of the year.

Tony Hayden writes to ask for a formula for the angle A north of east for the sunrise. It is hard to find, so here it is:

$$\sin A = \frac{\sin 23^\circ 27'}{\cos\,(\text{latitude})} \sin S,$$

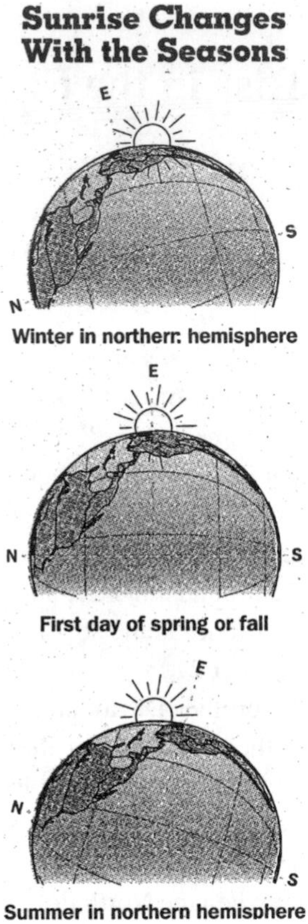

Sunrise Changes With the Seasons

E

S

N

Winter in northern hemisphere

E

N

S

First day of spring or fall

E

N

S

Summer in northern hemisphere

FIGURE 5 When the North Pole is tilted toward the sun, the sun rises north of east at every location on earth (*The Christian Science Monitor,* July 19, 1996).

where S is an angle measuring progress from the first day of spring (0° at about March 21, 90° at June 21, 180° at September 21, 270° at December 21). More such formulas and interesting questions appear in *100 Great Problems of Elementary Mathematics* by Heinrich Dörrie, Dover, 1965.

Charlie Conley sends in another intriguing question about the sun.

Question. My friend on Planet X reports that the sun stayed at the same spot on the horizon for a whole 24-hour day. How can this be?

Answer. Our friend certainly does not live on earth, where the sun never pauses in its noble journey across the sky. The only fixed star is Polaris,

FIGURE 6 The axis of Planet X is horizontal. When it points at the sun, the sun appears fixed in the sky as Planet X rotates.

the North Star, perched above the North Pole. *The north pole of Planet X must sometimes point at its sun.* Hence, Planet X's axis of rotation must be horizontal, not vertical, as in Figure 6. Then two days during Planet X's annual revolution around its sun, its axis points directly at or away from its sun, which appears to hang directly over the north or the south pole. If our friend lives on the equator, he sees the sun resting on the horizon.

Chuck Gahr remarks that this is almost exactly what happens on Uranus, which has a nearly horizontal axis of rotation. A resident on the Uranus equator sees the sun resting at one spot on the horizon during one of Uranus's 17-hour days, twice every orbit, which takes about 84 earth-years.

John Kruschke notes that although the sun remains on the same spot on the horizon for a day, Planet X's rotation does make the sun seem to rotate.

Kruschke also remarks that the seasons on Planet X are especially severe. At the summer solstice in its northern hemisphere, with its north pole pointing directly at the sun, the whole northern hemisphere has 24 hours of daylight. At this steamy North Pole, the relentless sun stays high in the sky all summer.

There is another answer to Conley's question, although not as good. Perhaps the same side of Planet X always faces its sun, much as the same side of the moon always faces the earth. Any resident in the circle of eternal twilight would see the sun forever stationary on the horizon. Of course, such a planet has an eternal day rather than what we meant by a 24-hour day.

Ruth Gatto points out that the Bible says our own sun once stood still at the Lord's command (Joshua 10:12–14).

We close with a final question from Conley.

Question. When and where on the earth are days 12 hours long?

Answer. On the equator every day is 12 hours long. In addition, there are two days of the year when every location has a 12-hour day, namely, the first day of spring and the first day of autumn. That is why those days are called the spring and autumnal *equinoxes* or "equal nights," i.e., 12-hour nights equal to 12-hour days.

EPISODE *3*

Leap Years

A young student "Tall" calls the TV show with a question that sounds deceptively straightforward but turns into one of our most popular topics.

Question. If you live for 6000 years, how many days will that be?

Answer. The first answer is (6000 years)(365 days/year) $= 2{,}190{,}000$ days. That is not quite right: it overlooks leap years. An older student takes leap years into account, adds $6000/4 = 1500$ leap year days, and comes up with 2,191,500 days. The answer is still not quite right. Every hundred years we skip a leap year (1900, although divisible by four, was not a leap year), so we subtract $6000/100 = 60$ days to get 2,191,440. The answer is still not quite right. Every four hundred years we put the leap year back in (2000 will be a leap year), so we add back 15 days to get 2,191,455, the final answer.

Leap years also make it hard to keep track of the day of the week.

Question. According to our current calendar, on what day of the week will January 1 fall in the year 3000? in the year 1,000,000? in the year $10^{1,000,000}$?

Answer. Throughout history famous "idiot savants" have been able to almost instantaneously identify the day of the week for any date. Their minds, shying away from most other normal functions, seem to develop an amazing familiarity with the calendar and its patterns. Professor John Conway of Princeton University has his easy "Doomsday" system for a normal person to perform similar feats, as described in his book *Winning Ways* with Berlekamp and Guy. Another method appears in *Mathemagics* by Art Benjamin. Our problem is a bit easier, however, in that the date is January 1.

In the thousand years from the year 2000 to the year 3000, there are 365,000 days, plus 250 leap days, minus 10 century leap days omitted, plus 3 leap days restored in the years 2000, 2400, and 2800, for a total of 365,243 days, or 57,177 weeks and 4 days. Since January 1 falls on a Saturday in the year 2000, it will fall four days later on a Wednesday in the year 3000.

To compute further into the future, note that the calendar has a 400-year cycle with the number of days equal to $(400)(365) + 100 - 4 + 1 =$ 146,097. By good fortune this is an even 20,871 weeks, so everything repeats every 400 years. Since 2000, 1,000,000, and $10^{1,000,000}$ are all multiples of 400, January 1 falls on the same day in all those years, namely, on Saturday.

Because of an accumulating astronomical error in our calendar, long before the year 1,000,000, New Year's Day would advance into summer. So we shall have to change the calendar, perhaps skipping some more leap years.

The current calendar with 146,097 days in 400 years averages 365.2425 days per year. The actual astronomical "tropical year," corresponding to our seasons, lasts only about 365.2422 days. This means we should skip a leap day in about $1/.0003$ years, or about 3000 years.

Good texts point out another effect: that the astronomical year itself is changing. First of all, the earth's revolution is gradually slowing down, about one day every 10 million years. (Then will the year become 364 days?) Amazingly enough, there is a much bigger effect in the opposite direction. Like a spinning top running down, as in Figure 7, the wobbling or precession of the earth's axis speeds up. Since our seasons are caused by whether the axis tilts toward or away from the sun, this increasing precession causes the tropical year to speed up, currently at about one day every 167,000 years. This would cause an accumulated error in the calendar of one day after about 600 years, around the year 2600. (After 600 years, the year has shortened .0036 days, the average year has been .0018 days shorter, accumulating in 600 years to about one full day shorter). Note that this coming effect is much greater than that due to the current mismatch between the calendar and the tropical year.

There is another big effect, which all texts I have seen seem to overlook: the *day* is getting longer, and longer days mean fewer days per year. The lengthening day requires the occasional addition of a "leap second" and the readjustment of all accurate clocks worldwide. Why is the day getting longer? Friction with the tides for example is causing the earth's

FIGURE 7 Like a spinning top running down, the wobbling or precession of the earth's axis speeds up and makes the tropical year get shorter.

rotation to slow, although there seem to be other complicating influences and this effect seems irregular and hard to measure. Ancient rocks that recorded daily lunar tides indicate that about a billion years ago the day lasted only about 18 hours (*Science News* 150, July 6, 1996, p. 4). If in a billion years the day lengthens by a factor of $\frac{4}{3}$, the number of days in a year goes down by $\frac{3}{4}$. In a billion years from now the year would lose about 91 days, or one day in about 11 million years. Actually, current controversial measurements suggest we are now losing about one day per 150,000 years, causing an accumulated error in the calendar of one day after about 550 years. (Combined with the previous effect of increasing precession, this yields a one-day error in about 400 years.)

Recently a controversy has raged on the internet over a peculiar definition usually used by astronomers. They mark the tropical year by the beginning of spring, while they probably should average over all the seasons. It makes a difference because the earth's orbit is not perfectly round, but a bit elliptical. Precession has a larger effect on the summer solstice (about June 21), when we are farther from the sun, and a smaller effect on the other seasons. As far as I can tell, the effects of this peculiar definition currently may be canceling out the other effects and keeping our calendar in almost perfect agreement with the tropical year, at least for the coming millennia.

Incidentally, the arrival of spring or the "spring equinox" is the moment when the earth is tilted at a right angle to the sun, not toward the sun as in the summer or away from the sun as in the winter. Consequently, night has the same length as daytime; hence the term spring "equinox" for "equal night." There is a certain variability in the arrival time of the spring equinox, listed in the table below. Ilan Vardi asks why the first day of spring in London sometimes come on March 20 and sometimes on March 21.

Spring Arrival ("Equinox")	Universal Time (London)
March 20, 1992	8:48 a.m.
March 20, 1993	2:41 p.m.
March 20, 1994	8:28 p.m.
March 21, 1995	2:14 a.m.
March 20, 1996	8:03 a.m.
March 20, 1997	1:55 p.m.
March 20, 1998	7:55 p.m.
March 21, 1999	1:46 a.m.
March 20, 2000	7:35 a.m.
March 20, 2001	1:31 p.m.
March 20, 2002	7:16 p.m.
March 21, 2003	1:00 a.m.
March 20, 2004	6:49 a.m.
March 20, 2005	12:34 p.m.

Source: U.S. Naval Observatory

The answer, of course, is "leap years." The extra day in a leap year, such as 1996, makes the first day of spring come about a day earlier (March 20, 1996) than the year before (March 21, 1995; see table). Since the year is actually about $365\frac{1}{4}$ days long, spring continues to come about $\frac{1}{4}$ day, or six hours, later every year, reaching March 21 in 1999, until the next leap year, 2000, brings it back to March 20.

The year is actually a bit less than $365\frac{1}{4}$ days long, by about 11 minutes, and this effect accumulates, until corrected by the omission of a century leap year, as in the year 1900. In the late 1800s, spring had fallen back to March 20, as it has now, but the omission of a leap year in 1900 advanced it back to March 21. (Many sources incorrectly still say that spring generally arrives on March 21.) Since 2000 is divisible by 400, it will be a leap year, and the next adjustment will not come until 2100. By then, spring will often be arriving on March 19.

The old Julian calendar did not omit any century leap years, and spring kept coming earlier and earlier. By the sixteenth century, the beginning of spring fell in early March. Pope Gregory XIII therefore created our current Gregorian calendar by excising 10 days from October 1582 and omitting the future century leap years not divisible by 400. Many European nations adopted the papal reform relatively quickly, although England and its Colonies, for example, held out until 1752. The current Gregorian calendar repeats every 400 years.

Careful examination of the table of spring equinoxes shows a certain irregularity in the advance from year to year. It averages about five hours and 49 minutes, but from 1997 to 1998 the spring equinox advances by six full hours.

Question. What causes the irregularities on the order of ten minutes in the time from one arrival of spring to the next?

Answer. Mike Bevan explains that the moon (and other planets), although much smaller than the sun, perturb the earth's orbit around the sun a bit. Although we think of the moon as circling the earth, the earth and moon actually both circle their "center of mass," although the earth is so much heavier than the moon that the center of mass lies inside the earth. Still, depending on which side the moon is on, the center of the earth can be either a bit ahead or behind the common center of mass, and spring arrives a bit early or late. Also, the axis of the earth wobbles a bit ("nutation"), causing the moment when the axis is at a right angle to the sun (the "spring equinox") to vary a bit.

Bob Swanson points out that at the spring equinox, day (sunrise to sunset) is a bit longer than night for two reasons. First, dawn comes before the center of the sun reaches the horizon, when just the top of the sun clears the horizon. Second, the earth's atmosphere bends sunlight so that we can see the sun even earlier. This all assumes no mountains get in the way!

We close with two final puzzles from the British *Games & Puzzles* magazine (No. 60, May, 1977), sent in by Edward Wallner.

Puzzle 1. Miguel de Cervantes and William Shakespeare both died on April 23, 1616. Who died first?

Answer. Cervantes of Spain, author of *Don Quixote*, died first. As a Catholic country, Spain followed the Gregorian calendar 10-day advance of 1582 immediately. Shakespeare's England, a Protestant country, did not implement the reform until 1752. Therefore, in 1616, April 23 occurred 10 days earlier for Cervantes than for Shakespeare.

Puzzle 2. The English archbishop Whitgift, first chairman of the committee that later produced the Authorized Version of the Bible, died on February 29, 1603. Explain how there could be a February 29 in a year not divisible by four.

Answer. In England in 1603 the year started on March 25, in the so-called Annunciation style. Therefore the day after March 24, 1603, was March 25, 1604. A leap day still occurred on February 29 as though the year had started on January 1.

EPISODE *4*

The Perfect Calendar

Our current calendar divides the 365-day year into 12 months, or 52 weeks plus one extra day. Could we not find a better calendar system? For example, are seven-day weeks best? Is there any way to avoid needing a new calendar every year?

Answer. You can use the same calendar every year (except leap years, which have two extra days) if you make the extra day a special holiday, perhaps at the beginning or end of the year, not part of any week. Chuck Gahr suggests calling it January 0. The popular proposed "World Calendar" of Figure 8, seriously considered by the League of Nations in the 1930s, inserts a "Worldsday" holiday at the end of the year, but keeps twelve 30- or 31-day months. The more radical proposed "Perpet-

FIGURE 8 The proposed World Calendar inserts a "Worldsday" holiday at the end of the year so that the next year will start on the same day of the week.

ual Calendar" has 13 identical 28-day months, so that not only can you use the same year's calendar every year, but also you can use the same month's calendar every month!

Bob Cohen sends in the even more radical "Secular Calendar," with six-day weeks. Jim Henry proposes 10-day weeks, with a third day off in the middle of the week. Mary Wright proposes five-day weeks with three 9.5-hour workdays. Dan Barbalace proposes five-day weeks with a one-day weekend called "Restday," following four workdays called Tuesday, Wednesday, Thursday, and Friday, "since nobody likes Monday." To compensate for the shorter weekend, he proposes shorter workdays and longer vacations.

One great virtue of all of these proposals is that businesses, governments, universities, and other institutions would not need to reschedule and recoordinate their activities at great cost every year.

As Cohen puts it, "everyone will need only two calendars: one for regular years and one for leap years. The two calendars could be combined by having a sliding door that can cover the leap day on non–leap years, like a pop-up book."

For more information see the web Home Page for Calendar Reform at http://ecuvax.cis.ecu.edu/%7Epymccart/calendar-reform.html.

Here is a final question to end on a lighter note.

Question. Everyone knows that February is the shortest month of the year. What is the longest month of the year?

Answer. At first it seems as if all the 31-day months tie. But in most locations, daylight saving time ends in October, making October an hour longer than the others. For second place, July and December sometimes have an extra "leap second" to compensate for the very gradual slowing of the earth's rotation.

EPISODE 5

Where Does the New Millennium Begin?

When does the new millennium begin? January 1, 2000? Actually it starts January 1, 2001. Our calendar was established to make the year 1 A.D. the first year after Jesus' birth, with the previous year 1 B.C. and no year 0. Hence 2000 is the 2000th year, the last year of the second millennium, and 2001 begins the third millennium.

The calendar actually got Jesus' birth too late by about six years; the exact date is still unknown. At his website, Michael Donner argues that there is one event from that period which we can pinpoint exactly, mathematically. What could it be? A lunar eclipse, which we can now compute occurred on March 12–13, 4 B.C. The contemporary historian Flavius Josephus tied such an eclipse to Herod's death, which we know certainly followed Jesus' birth. Adding 2000 years to 4 B.C. yields not 1996 but 1997, because there was no year 0. Therefore, as of March 13, 1997, we could be sure that it had been 2000 years since Jesus' birth and celebrate the arrival of the third millennium. So we may have missed it already!

Slightly different from the third millennium but still worth celebrating are the 2000s, which do arrive on January 1, 2000, as the 1900s end on December 31, 1999. This change in date is what causes the big problems with computers, which will interpret '00 as 1900 instead of 2000.

In any case, a bigger question remains. Assuming the third millennium arrives on January 1, 2001, *where* on Earth should the celebration begin?

You may have witnessed on TV on New Year's Eve the earlier celebration of the new year in other time zones. The most common answer is Greenwich, England, on the prime meridian, starting place of all time zones. Indeed, a plaque on the Old Royal Observatory announces, "The Millennium starts here" (Figure 9). But that is not the final word.

The year 2001, heralding the third millennium, will arrive earlier in England than in America, but it will arrive still earlier farther east in

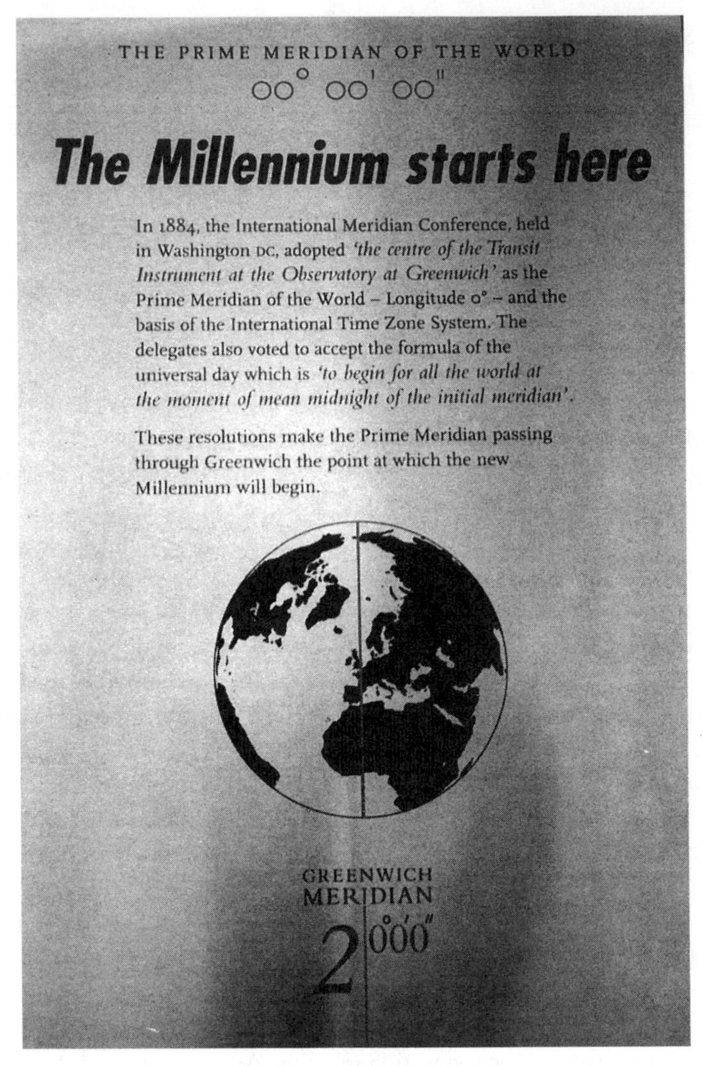

FIGURE 9 A plaque on the Old Royal Observatory in Greenwich, England, announces that "The Millennium starts here" (photograph by Bill Higgins).

Moscow, still earlier in Japan (that is why it is called the "Land of the Rising Sun"), and so on until you hit the International Date Line and drop back to the previous day. Figure 10 shows a map from *The Christian Science Monitor* labeled "Fiji First," showing Fiji right up against the 180-degree meridian of longitude. But that is not the final word either.

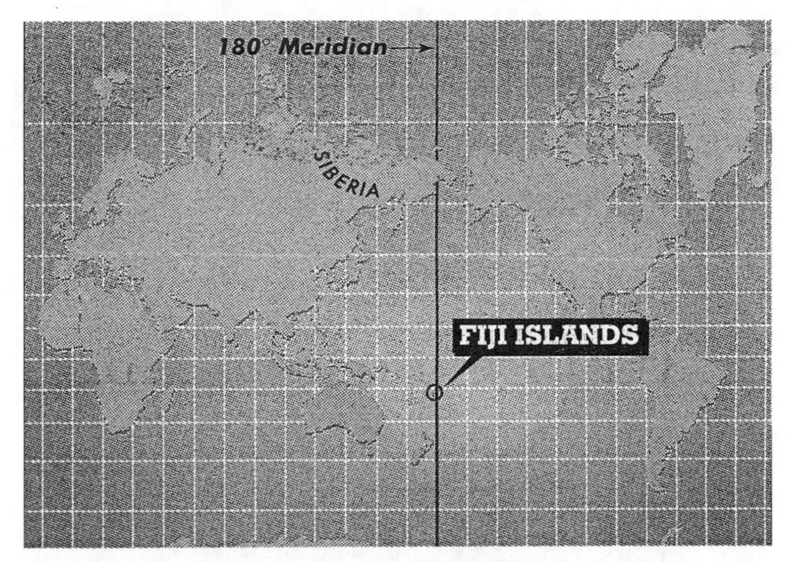

FIGURE 10 "FIJI FIRST: Island group, which sits on the 180-degree meridian along with Siberia, will be the first place to greet the dawn of the year 2000," wrote *The Christian Science Monitor* of March 18, 1998, but that is not the final word.

The dateline has long had an eastward bulge beyond the 180-degree line, including the South Pacific island of Tonga, situated in a later time zone. Tonga is closely followed by Chatham Island, in an odd time zone just 15 minutes behind. But that is not the final word either!

More recently President Teburoro Tito of the Kiribati Islands established a new claim to be first. This widespread island nation had been split by the dateline, with a different date in each half. Tito decreed a spectacular relocation of the dateline eastward around their boundary, as shown in Figure 11. Now Kiribati's Christmas Island will see the new millennium an hour before Tonga.

Aubrey Dunne suggests that the town of Anadyr in the Russian Far East might be another, if somewhat colder option. Determining the current time zone there turned out to be an interesting project. The best maps, including official US government documents, disagreed. Misha Chkhenkeli, a mathematical colleague from the former Soviet republic of Georgia, told me that there are many different times in Russia, including a universal "decreed time" occasionally announced from Moscow.

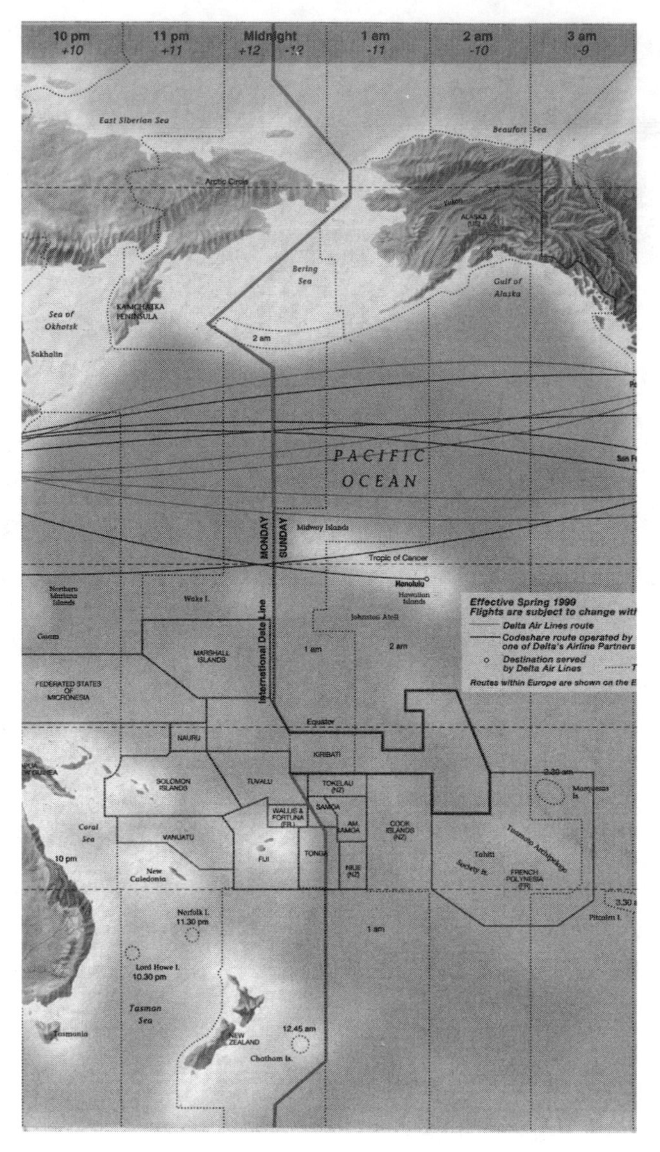

FIGURE 11 A spectacular relocation of the International Date Line eastward around the boundary of Kiribati makes it the first country of the new millennium. (Delta Air Lines.)

I call the Russian consulate in San Francisco, only to wait through a long recorded message, *in Russian.* Finally a voice answers, "Good afternoon," but those two words turn out to be the only English the good fellow knows. Eventually someone who speaks English appears and gives the incredible report that the time zone in the Russian Far East is just an hour or two earlier than San Francisco. So I decide to call instead the police department in Nome, Alaska. Information gives me the astonishing news that Nome has no police department, that the nearest police department listed is 500 miles away in Fairbanks. Finally I reach Sherry McBridge at the Chamber of Commerce in Nome, whose husband happens to work with radio broadcasting into the Russian Far East. Anadyr, it turns out, at 12 hours ahead of Greenwich, lags an hour behind Tonga and two hours behind Kiribati. Also, since New Year's Eve falls in the middle of winter in the Northern Hemisphere, Anadyr will lack the advantage of daylight saving time, which many Southern Hemisphere locations will employ, especially if it brings them the millennium first.

In any case, the first sizable city to see the new millennium arrive will be Auckland, New Zealand, and the last to see the old millennium go will be Honolulu, Hawaii.

Dave Gay proposes that we celebrate a new year not at midnight but instead at the dawn of the new day. He suggests that the celebration should take place in New Zealand on the mountain-top that receives the first light of the third millennium. Actually the day dawns sooner farther south, until at New Zealand's Scott Base on Antarctica, the sun is already up at midnight, although most research stations officially use Greenwich time.

Personally I hope to go to bed early and find a new millennium waiting for me when I awake.

Second Story: Probabilities and Possibilities

EPISODE 6

Baby Boys and Girls and World Population

Challenge. Suppose every couple keeps having children until they have a girl and then stops. Assuming boys and girls are equally likely, will this produce more baby boys or more baby girls in the whole population?

Answer. You might think more girls, since every couple has a girl. Or you might think more boys, since many a couple has lots of boys before the one girl. Actually, these two effects balance exactly, and one should expect equal numbers of boys and girls, as shown in Figure 12. Indeed, the first year couples expect half girls and half boys. The couples with boys try again, and the second year they expect half girls and half boys. Similarly, every year the remaining couples expect half girls and half boys. A shorter answer is that at every birth, no matter the history, a boy or a girl is equally likely, so one expects equal numbers of each.

It is astonishing to consider how the number of descendants grows over the years, as we did in the very first "Math Chat" newspaper column, which ran on Father's Day 1996. Suppose that a man has two children, say in the year 2000, that each of the children has two children 25 years later, that each of those children has two children 25 years later, and so on. In the year 3000, how many new descendants are born?

Well, the father has two children (first generation), $4 = 2 \times 2 = 2^2$ grandchildren (second generation), $8 = 2^3$ great-grandchildren (third generation), and in the year 3000 (forty-first generation), 2^{41} ever-so-great grandchildren. That's about 2.2 trillion, or 2.2 thousand billion, or more exactly

$$2,199,023,255,552$$

new descendants.

FIGURE 12 Every year produces half boys and half girls, as Aubrey Dunne illustrated with historic U.S. stamps. (Figure 12a)

But that answer seems too big. There would hardly be room on the earth for them all. And the whole population did not grow nearly that fast from the year 1000 to the year 2000. The current population of the earth is only about six billion. If *every* couple had two children, just to replace itself, the world population should not grow. So what is wrong?

The mistake comes from double counting, since after many centuries the descendants will run out of enough other people to marry and will have to marry each other. It's not fair to let them marry each other and have *four* children; if every couple in the world had four children, we would indeed run out of space.

If everyone had two children (say every 25-year generation), population would stay relatively constant, say for simplicity at $N = 3$ billion per generation. If descendants of our original father married other descendants or nondescendants at random (admittedly an oversimplification), the fraction f of the world population after n generations not descended from him would closely satisfy $f = (1 - 1/N)^{2n}$, falling below half the world population after $n = 31$ generations, or 775 years, when 52 percent of all newborns would be descendants of his, and reaching

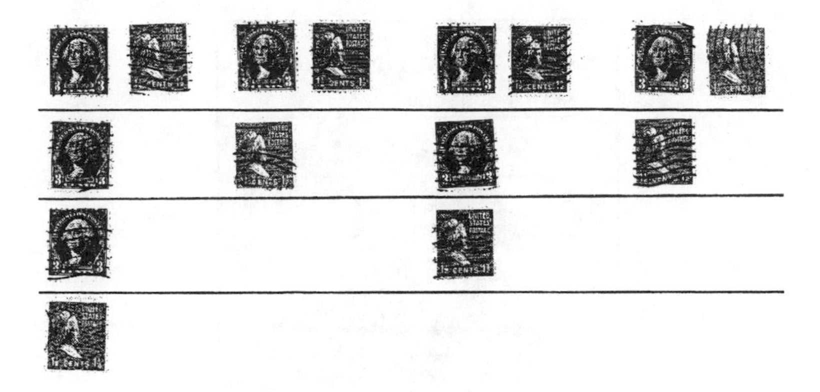

zero after $n = 36$ generations, or 900 years, when all newborns would be descendants of his! From the twenty-ninth century our original father hears a universal chorus from his three billion grandchildren inheriting the earth:

"Happy Father's Day!"

And that's how the first column ended. The rest of the columns' endings were less dramatic.

One sometimes hears that half the humans who ever lived are alive today. Can that be right?

Aubrey Dunne assumes that each generation is twice as large as the one before. So the previous generation was just $\frac{1}{2}$ as large as the current generation, the one before that just $\frac{1}{4}$ as large as the current generation, the one before that just $\frac{1}{8}$ as large as the current generation, and so on. Since $\frac{1}{2} + \frac{1}{4} + \frac{1}{8} + \frac{1}{16} + \cdots = 1$, all of the previous generations add up to just one current generation, and half of all humanity is alive today. Actually humanity has not always been growing that fast.

John Goekler cites a speculative scholarly article titled "How many people have lived on Earth" by Carl Haub in the February 1995 issue of *Population Today*, which finds much slower growth in earlier periods, as in Figure 13, and conjectures that only about five percent or so of humanity is alive today, out of a grand total of over 100 billion humans. Roughly a quarter have lived since 1400, roughly half since 1 A.D., and only about one billion before 8000 B.C. Modern Homo sapiens

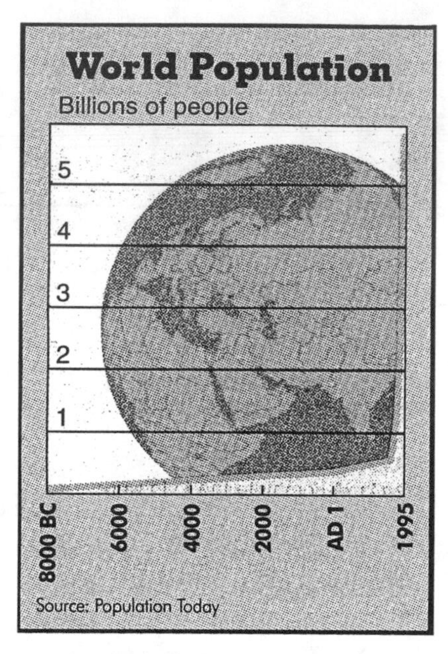

FIGURE 13 Although world population has exploded recently, only about five percent of all humans are alive today. (*The Christian Science Monitor,* January 31, 1997.)

may have appeared about 50,000 B.C. Edward Deevey, in "The Human Population" (*Scientific American*, September 1960), estimates 36 billion primitive Paleolithic hunters and gatherers going back a million years. Of course, the evidence is sparse and estimates vary widely.

EPISODE 7

Predicting the Random

Mathematical Trick.

1. What is $2 + 2$?
2. What is $4 + 4$?
3. What is $8 + 8$?
4. What is $16 + 16$?
5. Quick, pick a number between 12 and 5.

This trick has been circulating by email lately, and comes to us via Mary Lee. Your answer was 7, was it not? Try it on your friends.

Why do most folks pick 7? It would be a bit extreme to pick 12 or 5, or even 11 or 6. That leaves 7, 8, 9, and 10. Seven has great appeal. It is the number of days in the week, for example. Seven is special. The other numbers factor: $8 = 2 \times 4$, $9 = 3 \times 3$, and $10 = 2 \times 5$; but 7 is a "prime" number and cannot be factored into other numbers. Furthermore, the phrase "between 12 and 5" might somehow suggest $12 - 5 = 7$.

The preliminary addition questions are not important, except to put the respondent on edge and discourage concocting a deliberately unusual answer.

A similar old trick ends with "Name a vegetable," and invariably evokes the response, "Carrot!"

These tricks show that what appear to be "random" actions often follow hidden patterns or habits. The new mathematical theory of chaos, popularized by beautiful pictures of fractals and by the mathematician in the movie *Jurassic Park*, explains how some structure often underlies apparent disorder.

Here is a different kind of trick, in which the answer really is forced:

1. Pick a number from 1 to 10. Multiply by 9. Add the digits. Subtract 5.
2. Take that letter of the alphabet. (For example, if the result of step 1 is 3, take the third letter of the alphabet, C.)
3. Pick a European country that starts with that letter.
4. Pick an animal that starts with the last letter of your European country.
5. Pick a color that starts with the last letter of your animal.

Your color is orange!

Of course, the choice is essentially forced at each step. The first step always yields 4. The only European country that starts with the fourth letter of the alphabet, D, is Denmark. The familiar animal that starts with

FIGURE 14 The inescapable kangaroo.

K is kangaroo (although someone could pick koala). The familiar color that starts with O is orange.

If you try these tricks on friends, first hand them a sealed envelope with the answer, in order to produce the appropriate amazement when they open it after their response.

EPISODE *8*

The Bible Code and Personal Coincidences

Readers often ask about the best-seller *The Bible Code* by Michael Drosnin, which reports how computer decoding of the original Hebrew Bible predicts current events, such as the assassination of Yitzhak Rabin. My inclination is to attribute such discoveries to coincidence, not only because deliberate encoding sounds incredible, but also because the mathematics of coincidence can be very subtle.

Suppose your new phone number happened to be identical with the first seven digits of your U.S. Social Security number. You would have every right to be astonished. If either number were completely random, the odds of such a coincidence would be one in 10 million. Now suppose one hundred million of us compare our phone and Social Security numbers. With probability of over 99.99%, someone will have identical numbers. That is to be expected. Ask your friends: maybe we can find such a match. The point is that in a large context, some coincidences are bound to occur, and it is sometimes difficult to identify the appropriate context. If you look long and hard enough you will find a big coincidence. Williams statistician Dick De Veaux wonders how many ways of decoding the Bible have been attempted throughout history.

To test the hypothesis that in a big universe some unlikely things inevitably occur, I asked readers to send in true personal accounts. In the second best answer, Steve Gluck reports that "in a seemingly irresolvable dispute between my son and daughter, I proposed we flip a coin. My son said coin flipping was unfair because his sister always won. [After he chose tails,] thirteen consecutive flips were heads. The probability of such an event was 1 in 2^{13} or 8192." Actually, since such opportunities may have arisen on various, say eight, occasions, "Math Chat" puts the probability at about 1 in 1000.

In the best response, Fred Wedemeier reports that his grandfather lost his wedding ring working in the field on the family farm of Figure 15. One day Fred's father got down off the tractor and found the ring stuck in a small crack in one of the tractor's tires. Fred estimated that the probability that the crack on the tire would hit the ring in the field and that his father would get off at that time and find it to be 1 in 10^{22}. "Math Chat" estimates at least a 1 in 100 probability that the ring is favorably positioned at the surface, times a 1 in 100 probability that the tire groove will match up with the ring's location, times a 1 in 10 probability that the ring will lodge in the tire, times 100 days of opportunity over the years, times a 1 in 10 probability that someone would notice, for an overall probability of about 1 in 10,000.

FIGURE 15 Against odds of 1 in 10,000, a lucky farmer found a long-lost wedding ring in a tractor tire.

Charles Sullivan writes humorously, "I just flipped a coin 30 times, and got this sequence of heads and tails:

hthhhthhhtthtthhthhhtththhhth.

Amazing! The chances of getting that particular sequence are less than one in a billion! And yet it happened to me, sitting right here in my office, with an ordinary penny (1966). And on the first try!" The humor, of course, lies in the fact that there is nothing special about this sequence, and the probability of getting *some* sequence is 100 percent.

EPISODE 9

Incomparable Dice and Tic-Tac-Toe

S. H. Logue writes that the six faces of her four special dice of Figure 16 have these numbers on them:

A	1	1	5	5	5	5
B	4	4	4	4	4	4
C	3	3	3	3	7	7
D	2	2	2	6	6	6

You and your opponent each choose one die to use throughout the game. Each round you both roll, and the high number wins. Which is the best die to have, A, B, C, or D?

FIGURE 16 Which of these strange dice would you choose?

Timothy Dawn observes, "This game has the intriguing property that while D beats A, A beats B, and B beats C, C beats both D and A." Christopher Burns concludes that "If you get to choose a die already knowing what die your opponent has chosen, it becomes a sweet game—for you. If your opponent chooses A, you should choose D; if B, A; if

C, B; and if D, C." In each case you have a $\frac{2}{3}$ chance of winning. B beats C whenever C rolls a 3, which occurs $\frac{4}{6} = \frac{2}{3}$ of the time. C beats D the $\frac{1}{3}$ of the time that C rolls a 7, plus half of the $\frac{2}{3}$ of the time that C rolls a 3, for a total of $\frac{2}{3}$ of the time. Joan Carlson's eighth-graders actually constructed and rolled C against D 400 times and found C won 66.5 percent of the time.

Burns continues: "The game gets much more interesting if you don't know what die your opponent chooses. If your opponent chooses at random, pick C and you win 56 times out of 108. If you think your opponent is going to pick C, pick B. If you think your opponent is trying to think ahead of you, the trick is to know how many steps!"

Mathematical game theory says that the player who goes first has some "mixed strategy" (selecting various dice with various probabilities) that the player who goes second cannot beat. For this game, the safe mixed strategy is to use B and D at random (each with probability $\frac{1}{2}$). Now whether your opponent chooses A, B, C, or D, she wins just half the time. For developing this theory John Nash won the 1994 Nobel Prize in Economics. Understanding game theory more generally can help us understand international political and economic competition, for example.

Even tic-tac-toe was the subject of a serious mathematics seminar at the prestigious Institute for Advanced Study in Princeton on November 24, 1997. You may think the subject would not interest mathematicians, since the standard tic-tac-toe game between good players always ends in a tie. But what about playing in higher dimensions? If you play for 4 in a row in a $4 \times 4 \times 4$ cube (with 64 little boxes to put Xs in), the first player can always win. Professor Jozsef Beck of Rutgers University talked about going for 5 in a row in a $5 \times 5 \times 5 \times 5$ four-dimensional hypercube (with 625 little boxes) or, more generally, going for n in a row in a huge d-dimensional hypercube. He proved that even for d much bigger than n, the second player can sometimes force a tie.

Meanwhile, see the article on "Qubic: $4 \times 4 \times 4$ Tic-Tac-Toe" by Oren Patashnik in *Mathematics Magazine* (Vol. 53, 1980).

EPISODE *10*

Crossing a Rickety Bridge at Night

Conrad Weiser and Tiku Majumder send in the following question, supposedly used in Microsoft job interviews.

Microsoft Job Interview Question. "There are four people who need to cross a river at night as in Figure 17. There is a bridge that can only hold up to two people at a time. (It's rickety or something.) There is one

FIGURE 17 The famous Microsoft job interview question asks for the quickest way for the fast and slow to cross a rickety bridge in pairs.

flashlight that must be used when crossing. (It is extremely dark, and someone must bring the flashlight back to the others; no throwing anything, no halfway crosses, etc.). The four people take different amounts of time to cross the river. If two people cross together, they travel at the slower person's rate. The times are 10 minutes, 5 minutes, 2 minutes, and 1 minute for each of the four individuals." How fast can they complete the passage? Can they do it in 19 minutes? In 17 minutes?

Answer. They can certainly do it in 19 minutes, by having 1-minute "Speedy" escort the other three across the bridge. That's $2+5+10=17$ minutes crossing, plus Speedy's two 1-minute return trips, for a total of 19 minutes.

Amazingly enough, there is a 17-minute solution. First 1 and 2 cross, 1 returns, 5 and 10 cross, 2 returns, and finally 1 and 2 cross, for a total of $2+1+10+2+2=17$ minutes.

Princeton freshmen Amanda Fulmer, David Greco, Michael Lindahl, and Graham Meyer explain that this method wins essentially because $2-1<5-2$. If it took the second person 4 instead of 2 minutes, the best strategy would be to have Speedy escort the other three, because $4-1>5-4$. If it took the second person 3 instead of 2 minutes, the two methods tie, because $3-1=5-3$.

At a talk I give shortly afterwards, Alexander Klabin reports that "this question was asked of me during a recent interview I had at Goldman, Sachs & Co. . . . I am happy to say that I got the job at Goldman and I start in July!"

EPISODE *11*

Ideal Coinage

Have you ever noticed how many coins it sometimes takes to make change? In the United States, with our five coin denominations of 1¢, 5¢, 10¢, 25¢, and 50¢, it can take up to eight coins to make change up through 99¢. Indeed, 99¢ requires (1 × 50¢) + (1 × 25¢) + (2 × 10¢) + (4 × 1¢). Alice Loth wonders which different five denominations would minimize the number of coins ever needed to make change.

It turns out that with coins of 1¢, 3¢, 11¢, 27¢, and 34¢, as in Figure 18, you never need more than five coins to make change. For example, 99¢ = (2 × 34¢) + (1 × 27¢) + (1 × 3¢) + (1 × 1¢). Tenie Remmel reports that this is one of 1129 solutions, the one requiring the fewest coins on the average (3.343). Maybe we should adopt one of these systems, although they all make figuring out how to make change harder, especially since, as it turns out, you cannot always use the largest coin possible: a few middle-sized coins are sometimes better than one large coin and many small ones. For example, 54¢ = 2 × 27¢, but if you start out making change of 54¢ with a 34¢ coin, the remaining 20¢ requires 4 more coins: (1 × 11¢) + (3 × 3¢).

FIGURE 18 With coins of 1¢, 3¢, 11¢, 27¢, and 34¢, you never need more than five coins to make change.

EPISODE *12*

Infinitely Many Ping-Pong Balls

Jonathan Kravis and Dave McMath propose a paradoxical, one-hour process involving infinitely many Ping-Pong balls. In the first half-hour, two balls are placed in a huge empty barrel and then one is removed and discarded. In the next 15 minutes, two more balls are put in and then one of the three inside is removed. At each step, in half of the remaining time, two more balls are added and one is removed. How many balls are in the barrel at the end of the hour after infinitely many steps?

Answer. After n steps, there are n balls in the barrel. After infinitely many steps, there might be infinitely many in the barrel or there might be none at all or any number in between! Suppose for example that you label the balls 1, 2, 3, ... and that at the nth step you remove ball n. Then every ball gets removed at some time, and none are left. Of course, if at the nth step you remove ball $2n$, then balls 1, 3, 5, ... will never be removed. Infinite processes often defy intuition (Figure 19).

If you remove balls at random, it turns out that with 100-percent probability there will be no balls left. For example, the probability that ball one remains after one step is $\frac{1}{2}$, after two steps is $\frac{1}{2} \times \frac{2}{3} = \frac{1}{3}$, after

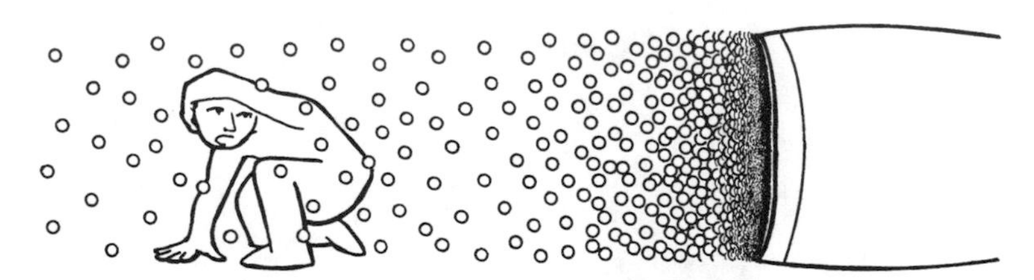

FIGURE 19 In mathematics, infinitely many Ping-Pong balls can accumulate or disappear in defiance of intuition.

three steps is $\frac{1}{2} \times \frac{2}{3} \times \frac{3}{4} = \frac{1}{4}$, after n steps is $1/(n+1)$, and after infinitely many steps is 0.

According to Jim Henry, for an engineer there will be only finitely many balls in the barrel, because after the nineteenth step, after 59 minutes and 59.99 seconds, the engineer will say the hour is up. For a quantum-relativist physicist there will also be only finitely many balls in the barrel, because the later ones would have to surpass the speed of light to get inside.

After this column appears, Luke Somers writes, "When I read the answer to the Ping-Pong ball challenge, I was astounded to see that part of it was simply wrong." Blair Perot writes, "Regarding the Ping-Pong balls, anything but an answer of infinity seems to be mathematical trickery.... It's not enough to say that eventually every ball could be removed (that is certainly correct), but only part of the equation. In the limit as every ball is removed—it is also replaced by two others."

"Math Chat" replies: If you rightly agree that in one scenario every ball is eventually removed (and replaced by two later ones), and you think that at the end infinitely many remain, I challenge you to name one ball that remains. Ball 1 is removed at the first step, ball 2 is removed at the second step, ball 3 is removed at the third step, and so on. Every ball is removed. No ball remains forever.

Here is a good analogy: the number of people who live forever is 0, even though the population of the world keeps increasing. By accelerating all of history into an hour, only the balls that "lived forever" remained.

This question comes to us from the popular text, *The Heart of Mathematics: An Invitation to Effective Thinking* by Edward Burger and Michael Starbird.

EPISODE *13*

Testing for AIDS

Chuck Gahr suggests that Batteries, Inc. knows that about one battery in a thousand is bad. They use a 99-percent accurate test on all their new batteries. If a battery tests bad, what is the chance it is bad?

Answer. Not 99 percent, but only about 9 percent. On the average, in a thousand batteries, the one bad battery usually tests bad, but of the 999 good batteries, 1 percent, or about 10, also test bad. So of the batteries that test bad, only about $\frac{1}{11}$, or about 9 percent, really are bad. There are many more good batteries, so that is where most of the mistakes occur. For the same reason, universal AIDS testing would initially yield false positive results for many more people than have the disease.

FIGURE 20 The laws of probability can make it dangerous to test the whole population for relatively rare diseases such as AIDS.

EPISODE *14*

Magician's Kings and Queens

Daniel Cebada says a magician takes a stack of eight cards (four kings and four queens, face down, in some order), turns the top one face up, moves the next one to the bottom of the stack, turns the next one face up, moves the next one to the bottom (below the card previously added to the bottom), and so on. To deal out an alternating sequence of four kings and four queens, how must you arrange the cards beforehand?

Answer. From the top, start with 2 kings, 1 queen, 2 kings, 3 queens. (Try it, then impress some friends.) You can find this answer by running the whole process backwards.

Suppose instead you have some number of cards, each with a letter of the alphabet. Can you arrange them to spell out one word or phrase, which when dealt out spell out another word or phrase? Brennie Morgan (yes, she is my mom, but she had no help from me) finds some five-letter examples: beats → baste, heats → haste, and peats → paste (not to mention error → error, as in Figure 21). Chuck Gahr finds ad rem → armed.

I offered a prize for an example with six or more letters. Sure enough, William Bond discovers bridge → big red and sawset → sweats. But the story does not end there. Reed Burkhart finds satrapy → stay rap (?) and Mr. Miller finds assesses → assesses.

Elementary school callers to my "Math Chat" TV show work on another puzzle. Setting $A = 1$, $B = 2$, $C = 3$, etc., find words that total 100. Annie reports TURKEY and TELEPHONE. Emily contributes STOVES. Colleen adds WRITING. Patchem Mortimer finds PUMPKIN. I do not know any others. Of course one could do a computer search.

FIGURE 21 Rearranging cards makes certain words mysteriously transform or reappear.

There are some interesting equations using the same numerical letter values:

$$\text{ARM} + \text{BEND} = \text{ELBOW};$$

$$\text{KING} + \text{CHAIR} = \text{THRONE}.$$

These may seem like amazing coincidences, but since the totals on each side are on the order of 100, the total on the right should have at least a $\frac{1}{100}$ chance of matching the total on the left, and therefore at least 1 percent of meaningful phrases should add up.

EPISODE *15*

Attic Lights, Marathons, and Fuse Timers

Attic lights. If you end up running up and down the stairs to get something as often as I do, you may like this question. Steve Jabloner says that inside his front door there are three on/off switches. One is attached to a regular light bulb on the third floor; the other two (like many in my house) are attached to nothing at all. You may work the three switches as often as you please. How many times do you need to visit the third floor to determine which switch is attached to the light bulb?

Answer. You can do it in one trip. Turn on the first switch, wait 10 minutes, and then turn off the first switch and turn on the second switch. Run upstairs. If the light is off but hot, it's the first switch. If the light is on, it's the second switch. If the light is off and cool, it's the third switch.

Erik Randolph creatively argues that, "There is no need to run upstairs at all: [just] watch for the blue spark, or even the buzzing sound [from] very, very slowly turning it on and off." His other methods involve a probe or compass.

Marathons. Of course, running up and down the stairs is good exercise and perhaps good preparation for this next question. Walter Wright's neighbor Joan was entered to run a special 26.5-mile marathon, and she hoped to average under nine minutes per mile over the total distance. As in Figure 22, she had a number of friends measure her time over various mile segments of the course, and for each mile that was measured, *in fact for each possible mile that could have been measured* (starting anywhere), her time was exactly nine minutes. Was she disappointed? NO—because she claims she met her goal of averaging *under* nine minutes per mile! Is this possible?

FIGURE 22 Joan ran each mile in nine minutes, but claimed to average *under* nine minutes per mile.

Answer. Yes, it is possible. For example, Joan could run the race in alternating four- and five-minute half-mile segments. Every mile would measure nine minutes. For example, if the mile starts 80% of the way through a four-minute segment, it would have 20% of that four-minute segment, the entire following five-minute segment, and 80% of the next four-minute segment, for a total of nine minutes. Nevertheless, since she starts and ends with a four-minute segment, there would be 27 four-minute segments and only 26 five-minute segments, and the average would be under nine minutes per mile. Indeed, the total time would be $(27 \times 4) + (26 \times 5) = 238$ minutes, for an average of $238/26.5 = 8.98$ minutes per mile. This can happen because the total distance is not an even number of miles.

It turns out that all that can be deduced mathematically is that she averaged over 8.83 minutes per mile (allowing unlimited speeds and neglecting any effects of Einstein's theory of special relativity). Indeed, alternating zero- and nine-minute half-mile segments would yield a total time of $(27 \times 0) + (26 \times 9) = 234$ minutes, for an average of 234/26.5, or about 8.83 minutes per mile.

It all goes to show how tricky averages and statistics can be. Walter Wright concludes, only somewhat facetiously, "In my work as a casualty actuary, the only difficult mathematical aspect is figuring out how to compute an average."

Fuse Timers. Presumably Joan's timers had good stopwatches, but suppose all you have are two one-hour fuses: lighting one end of a fuse will cause it to burn down to the other end in exactly one hour's time. You know nothing else about the fuses; in particular you don't know how long any segment of a fuse will burn, only that an entire fuse takes one hour. How can you tell when exactly 45 minutes have passed?

Answer. This is possible, but it takes two good ideas. The first good idea is to light both ends of a fuse at once. It will burn in a half hour, though it will not necessarily finish at the middle. The second good idea is to light one end of the other fuse at the same time, and its other end after a half hour (as measured by the first fuse). It will finish in 45 minutes: a half-hour at normal speed from one end and 15 minutes at double speed from both ends.

EPISODE *16*

Presidents' Names

Question. What happens when people pick letters at random? What is the chance that two people will randomly choose the same letter of the alphabet?

Answer. The chance that a second person will choose the same letter as the first is 1 in 26, or about 3.8 percent.

Question. What is the chance that the names of two random American presidents begin with the same letter? (See Figure 23.)

FIGURE **23** What is the chance that the names of two random American presidents begin with the same letter?

Answer. This question is a bit harder, since some letters occur more often than others as the first letter of American presidents' names. Three of the 42 presidents through Clinton have last names beginning with A (Adams, Adams, and Arthur), so the chance of two As is $(\frac{3}{42})(\frac{2}{41})$. Adding up such possibilities for all 26 letters yields a total of about 5.3

percent (or 5.1 percent if you count Cleveland just once). This is higher than 3.8 percent because some letters are more common than others, and that makes duplication easier.

Question. (Richard Thorne). Estimate the chance that at least 2 of the names of 10 random Americans will start with the same letter.

Answer. Erik Randolph first assumes that all letters are equally likely and computes the chance they are all different. The first letter can be any of the 26, the second any one of the remaining 25, the third any one of the remaining 24, and so on. The probability they are all different is

$$\frac{26}{26} \times \frac{25}{26} \times \frac{24}{26} \times \frac{23}{26} \times \frac{22}{26} \times \frac{21}{26} \times \frac{20}{26} \times \frac{19}{26} \times \frac{18}{26} \times \frac{17}{26},$$

or about 13.7 percent. Therefore the probability that at least two are the same is about 86.3 percent. Because some letters are more likely than others, the actual figure would be even higher.

Eric Brahinsky uses the last names of "Entertainment Personalities" in the 1997 *World Almanac,* of which a whopping 10.6 percent begin with an S. In 1609 random computer trials he finds that about 96 percent had duplicated first letters, much higher than the earlier 86.3 percent when all letters were equally likely. In such an experiment, statistics suggests you can be confident that the error is less than $1/\sqrt{1609}$, or about 2.5 percent. (This is the same $1/\sqrt{n}$ formula used to estimate errors in polls of n people: a future poll of 1609 people might announce that the percentage of Americans preferring Gore to Bush is 53 percent plus or minus 2.5 percent.)

Such random experiments can be useful in mathematics, too. I do not know a nice way to compute by hand the exact probability of duplication given the relative abundance of the initial letters.

EPISODE *17*

Presidential Elections

US Presidential Election Challenge Question. What is the fewest number of votes with which you could be elected President of the United States? Assume that there are just two candidates and say half the population in each state votes. Hint: you can win with fewer than half the votes.

Answer. Basically you need to win about half the votes in states with about half the electoral votes, or about 25 percent of the popular vote. Actually, since smaller states get more electoral votes per resident, you can win with 39 small states plus the District of Columbia and under 22 percent of the popular vote. So Dole could have beaten Clinton handily if he just could have redistributed his votes among the states. In 1888 Benjamin Harrison beat Grover Cleveland with fewer popular votes by winning lots of states by a few votes and losing big in the rest of the states.

If, contrary to the given stipulation, you consider the different voter turnout rates in different states, you could win with still fewer votes by winning states with low turnouts. In 1992, voter turnout varied from 42 percent of the voting-age population in Hawaii to over 70 percent in Maine, Minnesota, and Montana.

Question. What if there are three candidates?

Answer. With three candidates, a different argument says you need less than 0.1 percent of the vote, just enough to win Wyoming. If your opponents each fail to win a majority of electoral votes, the Constitution provides for the House of Representatives to choose among the three top electoral vote winners, including you, and if you can get the votes of

26 small state delegations in the House, you are elected! (Figure 24.) In 1824, Andrew Jackson won in popular and electoral votes, but failed to get a majority, and the House elected John Quincy Adams.

FIGURE 24 Is it possible to be elected President of the United States with less than 1% of the vote?

Third Story: Prime Numbers and Computing

EPISODE *18*

New Largest Prime Numbers

Most numbers can be factored: $24 = 3 \times 8, 25 = 5 \times 5, 26 = 2 \times 13$, but some numbers, called *prime* numbers, have no factors except themselves and 1. The first interesting example is 2, the *only* even prime, since all other even numbers are divisible by 2. The next primes are 3 and 5. The list of primes goes on forever:

$$3, 5, 7, 11, 13, 17, 19, 23, 29, 31, 37, 41, 43, 53, 57, 59, \ldots$$

You can start with any number, and keep factoring it until you have nothing but primes. For example, $210 = 6 \times 15 = 2 \times 3 \times 3 \times 5$. If 1 counted as a prime, you could add on lots of 1s: $210 = 1 \times 1 \times 1 \times 2 \times 3 \times 3 \times 5$, so mathematicians have officially decided not to count 1 as a prime, and then it turns out that every number has a unique breakdown into primes. Thus primes are the building blocks of all multiplication. They have fascinated mathematicians for centuries.

For a very big number n, it is hard to tell if it is prime or not. You could try dividing it by all smaller numbers, but if the number is really big, that takes too long, even for a big computer. So even though we know the primes go on forever, we do not know any *very* big ones.

In the past few years a number of bigger primes have been discovered. In September 1996, David Slowinski and Paul Gage at Silicon Graphics discovered a record-breaking prime on a Cray T94 supercomputer. Like most known big primes, it is a "Mersenne" number, obtained by multiplying 2 by itself a huge number of times and then subtracting 1. The new prime is

$$2^{1,257,787} - 1.$$

This prime is 378,632 digits long, enough to fill 100 single-spaced type-written pages. The previous record holder was $2^{859,433} - 1$, discovered in 1994. It is an open question whether infinitely many Mersenne numbers are prime, or even whether infinitely many Mersenne numbers are not prime.

Question. For this new prime number, what is the last digit? the second last digit? the first digit? the second digit?

Answer. The new prime starts with 41 and ends with 27. Fortunately, the last digit of powers of 2 repeats every 4 times and the last two digits repeat every 20 times. Dividing 1,257,787 by 20 leaves a remainder of 7. Since $2^7 - 1 = 127$, the last two digits are 27.

Finding the first two digits takes logarithms and a calculator. The log (base 10) of the new prime is just 1,257,787 times log 2, or about 378,631.615. Hence the new prime is about $10^{378,631.615} = 10^{.615} \times 10^{378,631} \approx 4.12 \times 10^{378,631}$, a number starting with 41 and having 378,632 digits.

Aubrey Dunne points out that this problem is much easier in base 2. In base 2, $2^{1,257,787}$ is just 1 followed by 1,237,787 zeros. To get the new prime, you subtract 1 (with lots of borrowing!) and get all 1s:

$$111111111111111111\ldots 1$$

(1,257,787 of them).

Just about two months later, in November 1996, a new largest known prime was found by Joel Armengaud of Paris:

$$2^{1,398,269} - 1.$$

Unlike the previous supercomputer discovery of September, it came on a personal computer, with help from many others in a huge project on the net. Likewise the next was discovered in August 1997 by Gordon Spence in 15 days of computing on his Pentium 100 PC:

$$2^{2,976,221} - 1.$$

This prime would take a solid month and over a mile of paper to write out.

FIGURE 25 Cal State student Roland Clarkson discovered a record-breaking prime number in January 1998.

A still larger prime was discovered in January 1998, by Roland Clarkson, a student at California State University Dominguez Hills, from Norwalk, California:

$$2^{3,021,377} - 1$$

(see Figure 25).

It has almost a million digits. Roland is now a student at Angelo State University in San Angelo, Texas, currently working on finding his next Mersenne prime.

The largest known prime at press time was discovered by Nayan Hajratwala on June 1, 1999:

$$2^{6,972,593} - 1.$$

The search for large primes provides a benchmark for our theoretical and computational progress. Perhaps the most important theoretical advance was the proof by Euclid (about 300 B.C.) that there are infinitely many primes. It is short and brilliant. It goes like this. Suppose there were just finitely many primes. Multiply them all together and add one. That would give a new number not divisible by any primes, a contradiction.

For example, suppose 2, 3, and 5 were the only primes. Multiplying them together and adding 1 yields 31, a new prime. Suppose 2, 3, 5, and 31 were the only primes. Multiplying them together and adding 1 yields $931 = 7 \times 7 \times 19$, yielding two new primes.

The Penguin Dictionary of Curious and Interesting Numbers by David Wells includes lots of fascinating prime numbers. One example is 82 81 80 79 78 . . . 1 (with all the numbers from 82 down to 1 in descending sequence). A second, 111 . . . 1 consists of 1031 repeating units 1; discovered by Williams and Dubner in 1986, it is the largest known repunit (repeating-unit) prime. A third prime has 1104 digits and all of them are prime. A fourth, 1999 . . . 9 consists of a one followed by 3020 nines.

EPISODE *19*

Four 4s

Howard Sheldon wonders how far you can get in forming numbers 0, 1, 2, 3, ... from four 4s and the standard mathematical operations $+$, $-$, $\times$, /, decimal point, square root, powers, and factorial (!). For example,

$$0 = 4 + 4 - 4 - 4 \qquad \text{and} \qquad 1 = 44/44$$

(Figure 26). Sheldon says he can get up to 30. How far can you get?

Answer. Eric Brahinsky, Dick Feren, Paul Goodrich, and Jim Bredt get

$$31 = 4! + (4! + 4)/4 \qquad \text{and} \qquad 32 = (4 \times 4) + (4 \times 4).$$

Roger Bliss, Robert Lewis, Mike Soskis, and Richard Thorne also get

$$33 = 4! + (4 - .4)/.4$$

and make it to 36. William Foster, William Hasek, and Michael Stern get

$$37 = 4! + \left(4! + \sqrt{4}\right)/\sqrt{4}$$

FIGURE 26 $1 = 44/44$, $2 = 4/4 + 4/4$, $3 = (4 + 4 + 4)/4$. How far can you get in forming numbers from four 4s?

and made it to 72. John Gordon gets 73 by bending the rules, using .4 with a bar over it to denote .444 . . . = 4/9 and noticing that

$$73 = \left(4! + 4! + \sqrt{4/9}\right) / \sqrt{4/9}.$$

If you start with 4 and keep pushing the square root key on your calculator, you get closer and closer to 1, so Carl Pomerance repeats that process infinitely many times and writes

$$73 = (\ldots\sqrt{\sqrt{\sqrt{4}}}) + 4! + 4! + 4!.$$

Harry Stern uses the mathematical symbol $\binom{24}{2}$ to denote the number of ways to choose 2 of 24 objects, which turns out to be 276, and writes

$$\binom{4!}{\sqrt{4}}/4 + 4 = 276/4 + 4 = 69 + 4 = 73.$$

I still do not know whether it is possible to get 73 without bending the rules.

Meanwhile Foster asks for the largest *prime* number you can get with four 4s. His best was

$$257 = 4^4 + 4/4.$$

A furious contest ensues. James Grimm and Michael Stern propose

$$577 = 4! \times 4! + 4/4.$$

Garrett Gray and Michael Eastep tender

$$331{,}777 = 4!^4 + 4/4.$$

Eric Brahinsky puts forward

$$479{,}001{,}599 = 12! - 1 = \left(4!/\sqrt{4}\right)! - 4/4,$$

which he finds in a list of primes in the *Penguin Dictionary of Curious and Interesting Numbers* by David Wells. Foster counters with

$$1{,}197{,}503{,}999 = 12!/.4 - 1 = \left(\left(4!/\sqrt{4}\right)! - .4\right)/.4,$$

which he confirms prime with his own computer program, with the fateful words, "I doubt that a larger one will be found by publishing date." The next day he finds on the net an algorithm for testing primes and verifies that the following champion is prime:

$$
\begin{aligned}
&167237572836228176439702951352259314668818658328450745257\\
&391783807540838931124074734171854131799624532703940947330\\
&352263813360214735203046497462027859741250149124013031423\\
&99999999999999999999999999999\\
&\quad = (5!! - .4)/.4 = \left(\left(\sqrt{4}/.4\right)!! - .4\right)/.4
\end{aligned}
$$

This "Elliptic Curves and Primality Proving" algorithm of Atkin and Morain (1993), which Foster uses, tests for primes as large as 10^{1000}. If you do not insist on logical 100-percent certainty, there are much faster probabilistic tests to verify primes with 99.99-percent or any desired level of certainty. Any one test may let certain "pseudo-primes" slip through. Professor Andrew Granville of the University of Georgia reports that, in a major advance, Jon Grantham's recent Ph.D. thesis showed that a dozen or so supposedly different kinds of pseudo-primes are really all examples of a single concept.

EPISODE 20

Powers of 5

Suppose you start with 5 and multiply it by itself thousands of times. What can you say about the answer, without attempting the horrific calculation? For example, Howard Sheldon asks for the remainder when you divide $5^{999,000}$ by 7.

Answer. The remainder is 1, as fourth-grader Madelyn Finucane (Figure 27) explains in Figures 28 and 29. The key pattern is that the remainder of powers of 5 after dividing by 7 repeats in cycles of 6.

FIGURE 27 Fourth grader Madelyn Finucane figures out that the remainder of $5^{999,000}$ divided by 7 is 1.

5409 Wild Turkey Lane
Columbia, MD 21044
June 22, 1997
mailed June 23
recd June 26

Dear Math Chat,

I tried the problem you put in the newspaper, and I think I got the answer. My grandfather gave me the article. He had cut it out of the newspaper from June 13th.

Here is how I got my answer. First, I did 5 to the first power, then I did 5 to the 2nd power all the way up to the 11th power. Then I divided all of them by 7 and wrote down the remainders. I saw a pattern in the remainders. The pattern of the remainders was 5, 4, 6, 2, 3, 1. (See page 1. All my multiplication and division are on pages 2-4.) My mom gave me a couple of numbers which were powers to find out the remainders . (See page 5.) For instance, 5 to the 42nd power divided by 7 has the same remainder as 5 to the 6th power divided by 7. To get my answer I divided 999,000 by 6 . There was no remainder so it went in the 6th row with 12, 18, 24, 30, 36, 42 and lots more. Being in that row means the remainder of 5 to the power divided by 7 is 1 (See page 1.) My answer to the problem is 1.

I'm 8 years old. I'm going to be in the fourth grade in the fall. Math and science are my 2 favorite subjects.

Sincerely,

Maddy

Madelyn Finucane

FIGURE 28 Finucane's explanation.

When you ignore multiples of 7 and look only at the remainder, you are doing arithmetic "modulo 7," as if $7 = 0$, $25 = 4$ (both 25 and 4 are 4 more than multiples of 7), and $699 = -1$. Modulo 7,

$$5^2 = 25 = 4, \quad 5^3 = 4 \times 5 = 20 = -1,$$

and

$$5^6 = 5^3 \times 5^3 = (-1) \times (-1) = 1.$$

Power	5 to the power	remainder of 5 to the power divide by 7 (1)
1	5	5
2	25	4
3	125	6
4	625	2
5	3,125	3
6	15,625	1
7	78,125	5
8	390,625	4
9	1,953,125	6
10	9,765,625	2
11	48,828,125	3
12		1

FIGURE 29 Finucane's calculations.

Because $999{,}000 = 6 \times 166{,}500$,

$$5^{999{,}000} = (5^6)^{166{,}500} = 1^{166{,}500} = 1.$$

It is no accident that powers modulo 7 repeat in cycles of 6. For *any* prime number p, powers repeat in cycles of $p - 1$. This important fact in number theory is called Fermat's Little Theorem, not to be confused with the famous Fermat's Last Theorem, whose recent proof after 350 years by Andrew Wiles got so much public attention.

After this answer appears in the newspaper column, I get some correspondence from Nick Drozdoff:

> I am a high school physics teacher, ex–electrical engineer and a current graduate student in theoretical/computational physics. [In your answer] were many equations such as $5^2 = 25 = 4$. This seems a peculiar form of mathematics. I don't personally see how $25 = 4$. What I can see is that 25 divided by 7 is three with a REMAINDER of four, but that is not what was written. What was written was $25 = 4$.
>
> ...High school kids do this in calculations all the time. They string their work together with equal signs skipping steps in favor of some sort of short hand. I drives me nuts because they have difficulty checking their work later. I encourage them to write each step of a calculation out. I discourage the tendency to string things together with equal signs that lead to visual/mathematical absurdities.
>
> Is this a common practice with math instruction? If so, I am appalled and shocked. I ask because this problem is not limited to any particular school. It's everywhere. I am concerned when I see this sort of thing in print in a place where people might regard such techniques as being accepted practice with experts, such as in *The Christian Science Monitor,* a respected newspaper not known for egregious mistakes. Please let me in on what this practice is all about.
>
> I do want you to know that I appreciate your column. How many newspapers have the guts to publish something that actually is interesting and requires or, rather, invites thought! Thanks for writing *Math Chat*.

All mathematics teachers know exactly what Drozdoff is talking about. Students who mean "$3x = 6$, therefore $x = 2$" instead incorrectly write "$3x = 6 = x = 2$." In contrast, we were not just stringing together different steps, but using the notion of "equality modulo 7" consistently throughout for numbers with the same remainder after dividing by 7, such as 25 and 4. To emphasize that this is a new kind of equivalence, mathematicians sometimes write $25 \equiv 4$ ("25 is congruent to 4") modulo 7 instead of $25 = 4$ modulo 7. On the other hand, mathematics needs to use many new systems with their own numbers and identities; for example, 25 and 4 could be considered different symbols for the same elements (or "congruence classes") of a new arithmetic modulo 7, where 25 really does equal 4.

Early number systems used just the positive integers

$$1, 2, 3, \ldots$$

Later number systems included negative numbers, then fractions, then imaginary numbers such as $\sqrt{-1}$ to get the so-called *complex numbers.* In 1843, to describe *rotations* in space, Sir W. R. Hamilton went a step further and invented the *quaternions.* Fifteen years later, he described the moment of discovery in a letter to Professor P. G. Tait:

> ... To-morrow will be the 15th birthday of the Quaternions. They started into life, or light, full grown, on the 16th of October, 1843, as I was walking with Lady Hamilton to Dublin, and came up to Brougham Bridge, which my boys have since called the Quaternion Bridge. That is to say, I then and there felt the galvanic circuit of thought close; and the sparks which fell from it were the fundamental equations between i, j, k; exactly such as I have used them ever since....

(from *Life of Sir William Rowan Hamilton* by Robert P. Graves, Volume II, Chapter XXVIII, http://www.maths.tcd.ie/pub/HistMath/People/Hamilton).

The quaternions come from adding three square roots of -1 to the real numbers: i, j, k. Now of course you have to know how to multiply quaternions. The rules are that

$$ij = k, \qquad jk = i, \qquad ki = j,$$

but

$$ji = -k, \qquad kj = -i, \qquad ik = -j.$$

Here reversing the order of multiplication changes the sign! Mathematicians would say that multiplication is not commutative.

Now it turns out that among the quaternions, there are infinitely many square roots of -1, not just i, $-i$, j, $-j$, k, and $-k$, but also, for example,

$$\frac{3}{5}i + \frac{4}{5}j,$$

because

$$\begin{aligned}\left(\frac{3}{5}i + \frac{4}{5}j\right)\left(\frac{3}{5}i + \frac{4}{5}j\right) &= \frac{9}{25}i^2 + \frac{12}{25}ij + \frac{12}{25}ji + \frac{16}{25}j^2 \\ &= \frac{9}{25}(-1) + \frac{12}{25}k + \frac{12}{25}(-k) + \frac{16}{25}(-1) \\ &= -\frac{9}{25} + 0 - \frac{16}{25} = -1.\end{aligned}$$

Similarly,

$$\begin{aligned}&\left(\frac{5}{13}i + \frac{12}{13}j\right)\left(\frac{5}{13}i + \frac{12}{13}j\right) \\ &\quad = \frac{25}{169}i^2 + \frac{60}{169}ij + \frac{60}{169}ji + \frac{144}{169}j^2 \\ &\quad = \frac{25}{169}(-1) + \frac{60}{169}k + \frac{60}{169}(-k) + \frac{144}{169}(-1) \\ &\quad = -\frac{25}{169} + 0 - \frac{144}{169} = -1.\end{aligned}$$

In fact, for any triple of natural numbers such that $a^2 + b^2 = c^2$,

$$\frac{a}{c}i + \frac{b}{c}j$$

is a square root of -1.

Towers of 5 Challenge. Returning to powers of 5, William Foster sends a harder challenge involving towers of powers of 5, such as

$$5^{5^5} = 5^{3125}$$

(not 3125^5; you have to start from the top). Foster asks for the remainder after dividing by 7 of

$$5^{5^{5^{\cdots 5}}} \qquad \text{(a tower of 999,000 5's).}$$

Answer. Michael Stern uses the computer program *Mathematica* to compute that 5^5, 5^{5^5}, and $5^{5^{5^5}}$ all leave remainder 3 when divided by 7, and surmises correctly that the simple pattern continues. Here is why. Consider the example $5^{5^{5^5}}$. Since as discussed above powers of 5 modulo 7 repeat in cycles of 6, we need to consider 5^{5^5} modulo 6. In a similar fashion, powers of 5 modulo 6 repeat in cycles of 2. Because any power of 5 is odd, therefore, modulo 6, $5^{5^5} = 5^1 = 5$. Finally, modulo 7, $5^{5^{5^5}} = 5^5 = 3125 = 3$, as claimed. Since the third level on up is always a power of 5 and hence odd, this argument applies to towers of all heights.

We conclude this episode with two questions about factorials. Given any number such as 5, you compute its factorial, denoted 5!, by multiplying it by all smaller numbers:

$$5! = 5 \times 4 \times 3 \times 2 \times 1 = 120.$$

Note that 5! ends in a zero.

Challenge (Howard Sheldon). Determine the number of zeros at the end of 25! (What about 5^{10}!?)

Answer. 25! ends in six zeros. To understand why, start with the easier case of $10! = 10 \times 9 \times 8 \times 7 \times 6 \times 5 \times 4 \times 3 \times 2 \times 1 = 3{,}628{,}800$. It ends in two zeros, one because of the factor of 10 and one because of the separate factors 5×2. Since 2s are much more common than 5s, it is the number of factors of 5 that determines the number of terminal zeros. In 25!, $25/5 = 5$ of the factors are divisible by 5, and one (namely 25) is divisible by 5 twice, for a total of 6 terminal zeros.

Advanced Challenge. Determine the number of zeros at the end of 5^{10}!

Answer. 5^{10}! ends in 2,441,406 zeros. In 5^{10}!, $5^{10}/5 = 5^9$ factors are divisible by 5, $5^{10}/5^2 = 5^8$ are divisible by 5 twice, $5^{10}/5^3 = 5^7$ are divisible by 5 three times, and so on, for a total T of

$$T = 5^9 + 5^8 + 5^7 + \cdots + 5^1 + 1.$$

To compute T easily, note that

$$5T = 5^{10} + 5^9 + 5^8 + \cdots + 5^1.$$

When you subtract T from $5T$, most terms cancel, and you get

$$4T = 5^{10} - 1$$

and

$$T = (5^{10} - 1)/4 = 2{,}441{,}406.$$

In general, $5^N!$ has $(5^N - 1)/4$ terminal zeros.

EPISODE *21*

The 2000 Census

Gordon Squire notes that planners for the 2000 US Census wonder what to do about missing certain groups of people such as many homeless in New York City. Suggestions include statistical random sampling. What approach would you suggest?

Answers. Two interesting suggestions come from students calling the "Math Chat" cable TV show in Williamstown, Massachusetts. Second-grader Nicholas De Veaux proposes basing the census on the phone book. Of course there is usually just one name per family, some phones are not listed, and some folks (such as the homeless) have no phones. Third-grader Eli Raffeld suggests announcing that anyone could stop in at town hall, register family and friends, and get $10 per registration. For the United States, which has a population of about 260 million, that could cost $2.6 billion, which is almost exactly what the census did cost in 1990.

Williams College student Heath Dill reasons that since he is one person in a 100-square-foot room, and the area of the United States is about 100 trillion square feet, the population of the United States must be about a trillion (1000 billion)!!

Robert Lewis, who actually spent two years homeless in New York City "on the run" from forgery charges, writes (from prison): "I can tell you one thing homeless persons all have in common. They eat." He suggests requiring census registration in order to eat at food distribution centers.

The census reminds me of an old question, recently submitted by Aubrey Dunne, Michael Marcotty, and Dave Rossum, which at first seems to lack enough information to solve the problem.

Census taker: "How old are your three daughters?"

Mrs. S: "The product of their ages is 36, and the sum of their ages is the address on our door here."

Census taker: "I'm good at math, but I cannot tell."

Mrs. S: "My eldest daughter has red hair."

Census taker: "Oh thanks, now I know."

Can you figure out how old the three daughters are?

Answer. Since the census taker cannot tell their ages from the product and sum, there must be at least two possibilities with product 36 and the same sum. Trial and error yield just 1, 6, 6 and 2, 2, 9, both of which have sum 13 (which must therefore be the address as in Figure 30). Reference to an "eldest" daughter rules out the first possibility and means that there is a 9-year-old eldest daughter and 2-year-old twins.

FIGURE 30 How can the census taker figure out the daughters' ages?

Noelle Matteson (age 11) notes the crucial insight, that the census taker knows the address even though we the readers do not. James Fahs and Diane Larrabee note that the hair color is a "red hairing."

Some readers argue that one of two 6-year-olds could be "older" by 11 months, or 11 days if adopted, or even 11 minutes if twins, so you cannot rule out 1, 6, 6 for sure.

EPISODE 22

Can a Computer Have Free Will?

Is it logically possible for a computer to have free will?

Seth Rogers argues that "since a computer must follow its program it does not have choice. Many contend that the brain works the same way by firing neurons, but they do not distinguish the mind from the brain. Since a computer and a brain perform similar functions, the study of AI [artificial intelligence] can reveal the boundary between the brain and the mind."

On the other hand, Michael Jackson argues that if people can have free will, so can computers. He predicts that some day computers will pass the "Turing Test" of successfully posing as humans, say in a chat room on the internet, with similar intelligence, humor, and unpredictability. Luis Baars quips that his computer certainly has a will of its own.

"Free will" classically means that one's actions are not predetermined. It is well recognized that computers often come up with unexpected responses and solutions. Whether computers are in some deeper sense predetermined is a harder question. In modern physics, quantum mechanics says that nothing is predetermined: that almost anything has some (perhaps very tiny) probability of occurring.

Deb Bergstrand suggests that free will should mean the freedom to do what one decides to do, and Ruth Gatto points out that computers "operate freely where the human mind cannot cope with massive amounts of technical information."

Meanwhile, Hymro Schnepple cites philosophers from Plato to Spinoza (free will as an illusion) to Nick Herbert (who describes the discovery process as "the porridge wakes up") and concludes that computers cannot have free will, because it "exists only in transcendence—perceived through the higher Atman, or Soul-body, which physical-universe computers simply do not have."

Gatto concludes that "through the computer world, mankind for the first time can see its mental state objectively. Hopefully this will guide us to a higher awareness of the infinity and possibilities of the mind's intellect."

Artificial and human intelligence appear in a recent list by the famous mathematician Steve Smale of 18 important mathematics problems for the next century (*The Mathematical Intelligencer,* Spring 1998). Number 1 is the notorious Riemann Hypothesis, related to the distribution of prime numbers. Number 18 asks,

> "What are the limits of intelligence, both artificial and human?"

In the best answer submitted, John Robertson argues that there are indeed limits to artificial and human intelligence, since the physical universe itself is finite, but that these limits are quite high, as illustrated for example by mathematics. Then he asks the big question: "Will we solve the important problems that mankind faces? These are, as is well known (and listed here in apparent increasing order of difficulty), the question of why is there something instead of nothing, the nature of consciousness, the generalized Riemann hypothesis [Smale's number 1 question, related to prime numbers], and the question of whether there are (nontrivial) 3×3 magic squares such that every entry is a perfect square." (The final question, however difficult, is probably not the most important.)

Along the way, Robertson argues that artificial (machine) intelligence is at least as great as human intelligence, because the brain is a machine. On the other hand, he thinks that brains probably can be genetically engineered to do anything a futuristic computer can do. Mike Jackson wonders why we tend to evaluate artificial intelligence by comparing it to human intelligence in the first place.

Modern mathematics proves results from generally accepted axioms. In 1931, the mathematician Kurt Gödel astonished the world by proving roughly that for any such system, if it is self-consistent, there are truths that cannot be proved. This result dashed all hopes of mathematically deriving all truth from a fundamental set of self-evident axioms.

Why does our mathematics work so well in explaining the universe? Might alien cultures have different mathematics from ours? These intriguing questions appeared in an article, "Useful invention or absolute truth: what is math?" by George Johnson ("Science Times" section of *The New York Times,* February 10, 1998). A graphic (credited to Dr.

George Lakoff of the University of California at Berkeley) gave three highly debatable reasons why an alien culture might not understand our number π (the ratio of the circumference of a circle to its diameter, about 3.1416).

1. The aliens might have a different number system. If they have four hands and 24 fingers as in Figure 31, they might use a number system based on 24 rather than 10, with 23 different symbols such as 1, 2, 3, 4, 5, 6, 7, 8, 9, A, B, C, ... before using the symbol 10 for twenty-four. In the two systems, π would be written out in completely different ways. On the other hand, our mathematicians and theirs could still recognize in each other's different symbols the same underlying number value π.

FIGURE 31 Even an alien with 24 fingers, a curvy desktop, and a crazy ruler would still discover the fundamental constant π.

2. The aliens might work not on flat desktops but on curvy surfaces such as spheres, where the ratio of the circumference of a circle to its diameter is variable and less than π. True, but for very small such circles the ratio approaches π, which is still an important number in such "spherical geometry."

3. The aliens might for example measure distances with some crazy scale in which two inches are more than twice as long as one inch, and four inches are more than twice as long as two inches, and so on. True, and we ourselves use "logarithmic" and other scales sometimes, but the usual scale is the very natural and special one in which $1 + 1 = 2$ and $2 + 2 = 4$ and so on. Their mathematicians would be sure to know and love this scale and our favorite constant π.

Of course, even if the substance of mathematics is similar in different cultures, the language and the symbols are different. The Western world currently uses the "Arabic" number symbols 1, 2, 3, etc., while Arab countries today use "Indic" numerals like those in Figure 32. Our "Arabic" number symbols, although superior to Roman numerals, have some disadvantages. The 4 and the 9 can sometimes be confused. The 5 takes two pen strokes. The 1 can be confused with the letters "I" and "l." The zero can be confused with the letter "O."

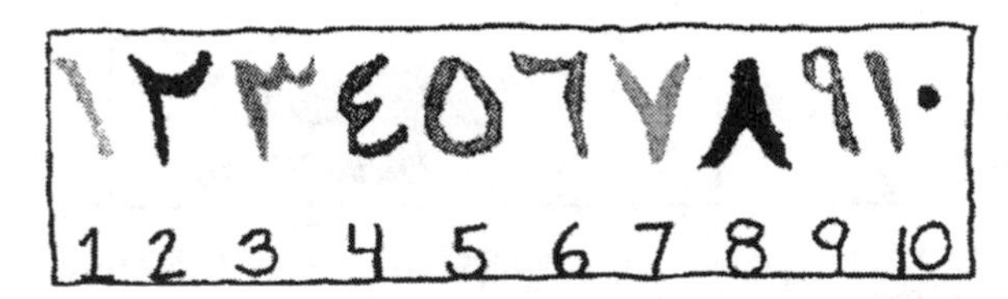

FIGURE 32 While we use Arabic number symbols, Arab countries use these Indic numerals. (*The Christian Science Monitor*, July 9, 1998.)

Jim Henry asks for a completely *new* set of symbols for the numbers from 0 to 9. He adds, "We can aim at having a Year 2000 switch-over to [the best] new, improved way of writing numbers."

Three winning answers are pictured in Figure 33. Bill Hasek's 1 has one stroke, 2 has two strokes, 3 has three strokes, 4 has four strokes, 5 resembles an F for five, and the rest are curvy, including an upside-down empty bowl for 0. Michael Marcotty's, although similar to our current number symbols, have a neat feature: the symbol for 1 has one angle to it, the symbol for 2 has two angles to it, and so on. Marcotty further proposes replacing the decimal point by the European decimal comma, to avoid confusion with his new raised dot for zero. Dave Rossum's "clock" numerals are original, systematic, and easy to write, although they could be confused, especially on dice or other objects which might be upside down. (Erik Randolph mentions the trouble his young son had confusing M and W, for example.) Incidentally, John Robertson mentions a disadvantage of the English pronunciation of letters as compared with numerals: one often needs to say "b as in boy," but not "9 as in. . . ."

FIGURE 33 Readers propose new number symbols. (*The Christian Science Monitor,* July 9, 1998.)

After announcing the preceding suggestions, we received two marvelous different explanations of the origin of our number symbols. Louis Hansell describes how perfectly the symbols model natural hand signals for numbers: for example, three horizontal fingers look like the symbol 3; a clenched fist plus a thumb up look like the symbol 6. Calvin Senning sends a 30-year-old clipping from *Reader's Digest* which says that a Moroccan genius over a thousand years ago designed the symbols to have the appropriate number of angles, like the ones submitted by Michael Marcotty.

Fourth Story: Geometry

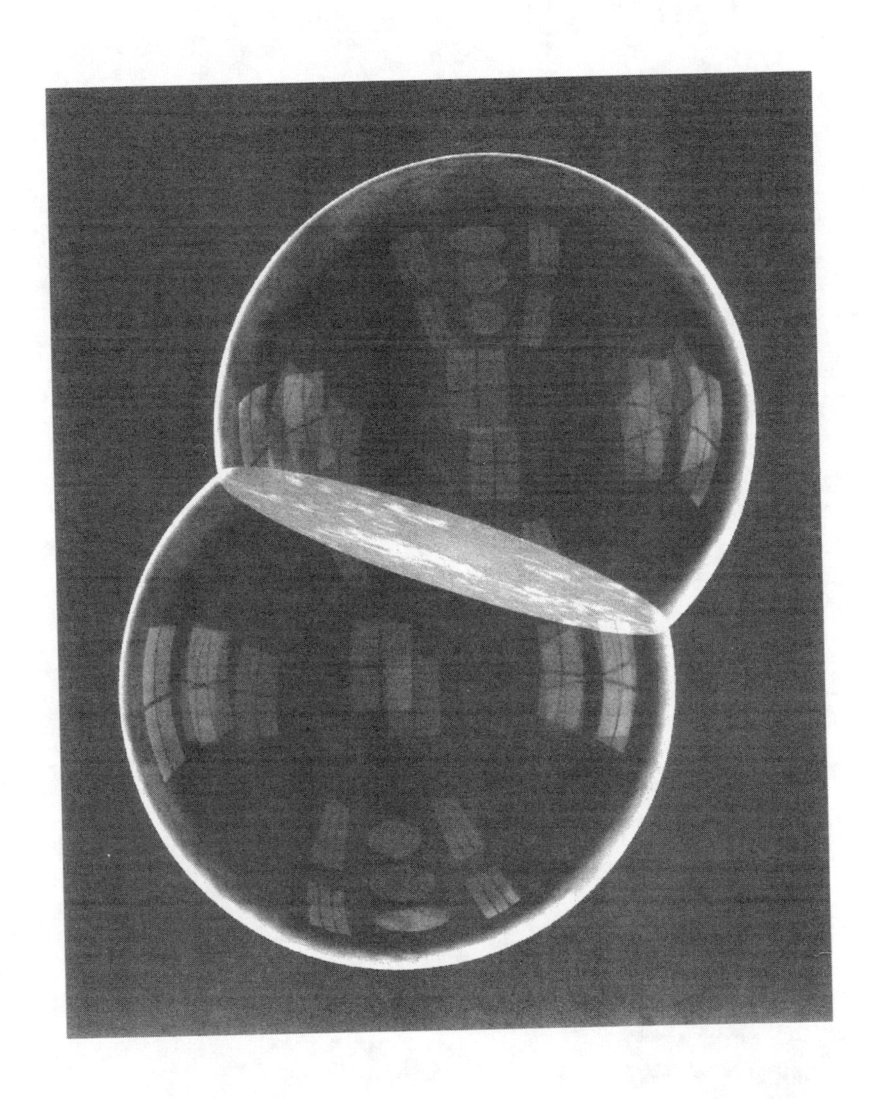

EPISODE 23

The Double Soap Bubble Breakthrough

The Double Soap Bubble Breakthrough of 1995, ranked by *Encyclopedia Britannica* as second only to Wiles's proof of Fermat's Last Theorem, shows how close the frontiers of mathematics can be to college mathematics and calculus.

If you never know anything else about calculus, you should know that the most important topic is getting the maximum (or the minimum) out of something: maximizing profit, minimizing time. Some might say that's what life's all about: maximizing happiness, minimizing hassle. The unit is usually called "Maxima–Minima Problems" and often features notorious word problems. The geometry problems are good ones, because it's easier to picture what's going on. Here is one of the standard questions. Maybe you can guess the answer.

Question 1. Find the rectangular enclosure of given area which uses the least amount of fencing.

Answer. The square (see Figure 34).

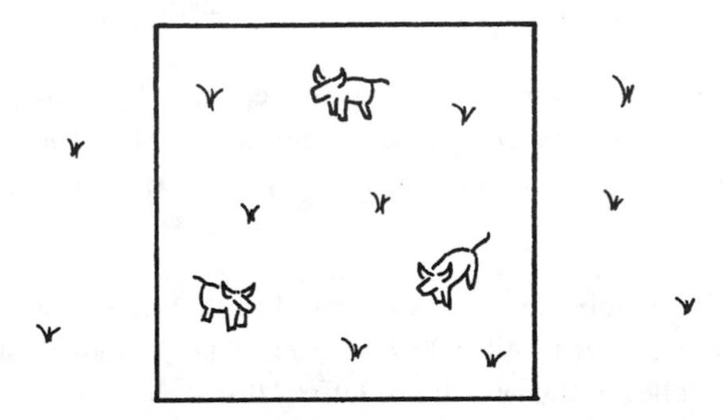

FIGURE 34 The square encloses more area with less fencing than any other rectangle.

This can be shown handily as an illustration of calculus. The trouble is that the answer looks obvious, so why all the fuss?

Question 2. Find the most efficient shape for two identical rectangular pens, adjacent along their length.

Answer. Not two squares, but two rectangles that are a third longer than wide, as in Figure 35. Why? To take fuller advantage of the common wall that serves both. This happens to be almost as easy a calculus problem, with an answer that few students would guess.

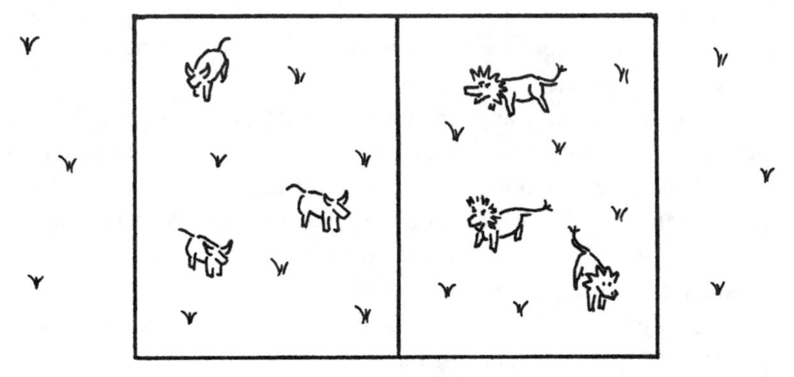

FIGURE 35 The most efficient rectangles enclosing two identical areas are a third longer than wide, to take fuller advantage of the common wall which serves them both.

New Double Pen Question. What is the most efficient shape for two identical adjacent pens if any shapes are allowed, not just rectangles?

Answer. Two overlapping circular pens, separated by a straight line as in Figure 36. The three fences meet at equal angles of 120 degrees. (This shape does better than one circle divided by a long straight line down the middle.)

You might think this fact was proved by the ancient Greeks. Actually, it was not proved until 1990, by a group of undergraduate students at Williams College: Manuel Alfaro, Jeffrey Brock, Joel Foisy, Nickelous Hodges, and Jason Zimba. Their paper was published in the *Pacific Journal of Mathematics* in 1993.

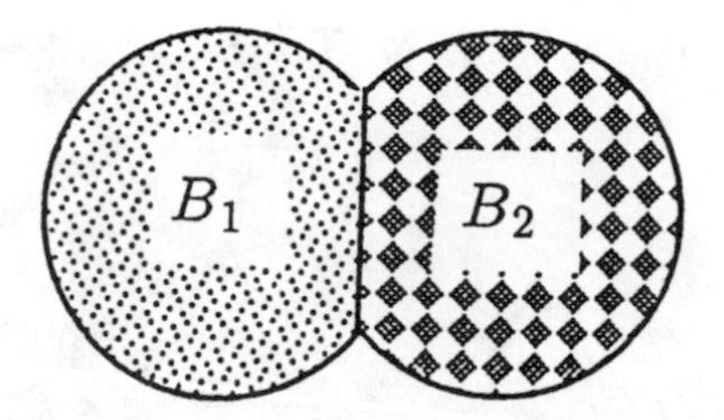

FIGURE 36 The most efficient shape for two identical adjacent pens if any shapes are allowed consists of two overlapping circular pens, as proved in 1993 by undergraduates (illustration from their paper, "The standard double soap bubble in $\mathbf{R}^2$ uniquely minimizes perimeter," *Pacific J. Math.* (1993)).

This double pen looks like the familiar double soap bubble of Figure 37, in which the bubbles also always meet at angles of 120 degrees. Every double soap bubble, whether or not the two bubbles are the same size, is probably a most efficient shape for enclosing those two given volumes of air in space, but mathematicians, despite much effort, have been unable to decide for sure. Work in the 1990s uncovered some weird alternative shapes, like the crazy double bubble computer simulation of Figure 38, in which the second region wraps around the first like an inner tube around a beach ball. So far these new ones have turned out to be less efficient and too unstable to be made as physical soap bubbles.

In a major breakthrough in 1995, Joel Hass of the University of California at Davis, Michael Hutchings, currently of Stanford, and Roger Schlafly, president of Real Software, Santa Cruz, California, announced a computer proof that the familiar double soap bubble is best for our main case when the two bubbles are the same size, as pictured in Figure 37. The proof ran on an ordinary PC in about 20 minutes. Hass and Schlafly got their inspiration during a calm stretch between rapids while kayaking down the south fork of the American River. Their work depended on the earlier work of the undergraduates. Indeed, Hutchings got started as an undergraduate at the Williams research site.

Last-minute news flash. In June, 1999, Professor Thomas Hales of the University of Michigan announced a proof for the case of not two but infinitely many regions of unit area, not in space but in the plane. Here the solution is a tiling by regular hexagons. This may partly explain why hexagons occur in nature, as in the bees' honeycomb. For more information see Math Chat of June 17 at www.maa.org or the Hales web page at www.math.lsa.umich.edu/~hales/.

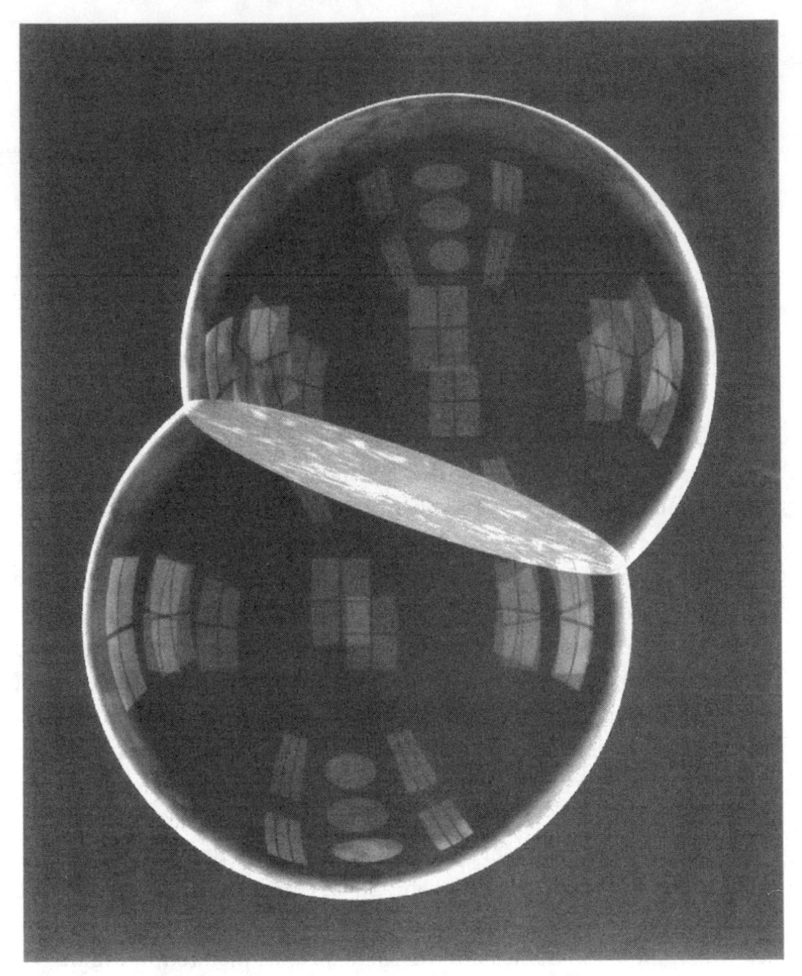

FIGURE 37 The double soap bubble, most efficient enclosure? (computer simulation by John M. Sullivan, University of Illinois at Urbana-Champaign).

Last-second news flash. As this book goes to press, a proof of the full Double Bubble Conjecture, that the familiar double soap bubble is best for any two volumes in space, seems to have been found by Hutchings, Morgan, Manuel Ritoré, and Antonio Ros (the latter two of the University of Granada). The proof shows that other contenders are unstable, without the use of computers. Moreover, Morgan's NSF "SMALL"

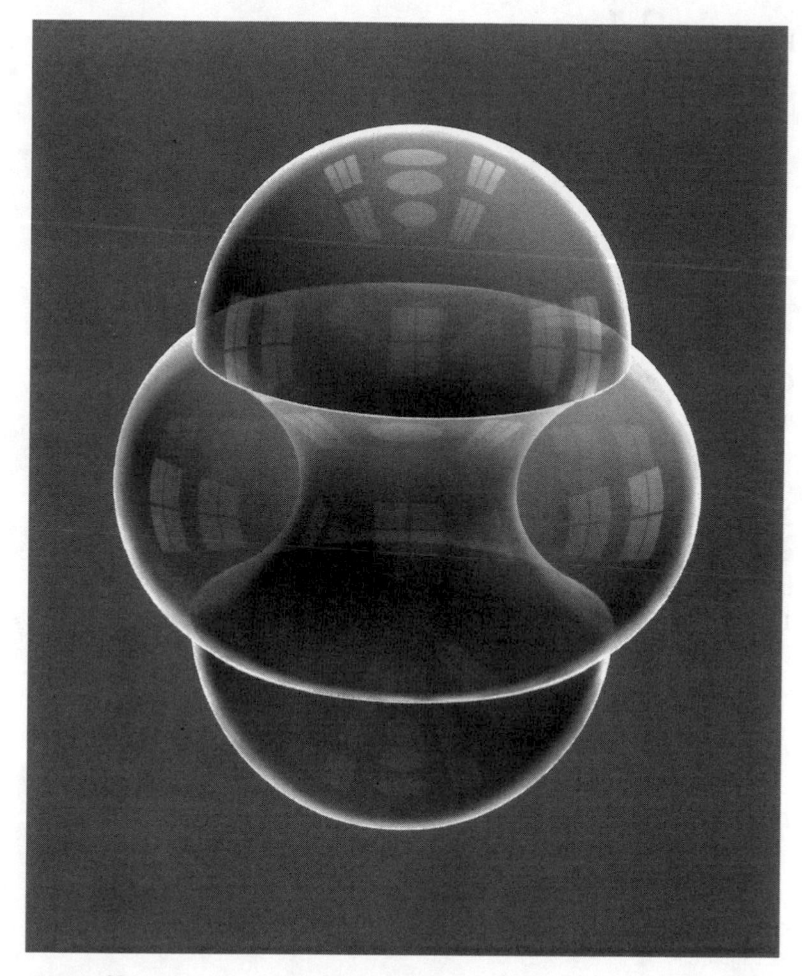

FIGURE 38 A strange alternative double bubble with an innertube bubble around a central peanut-shaped bubble (computer simulation by John M. Sullivan).

undergraduate research Geometry Group, consisting of Cory Heilmann (Williams College), Yvonne Lai (MIT), Benjamin Reichardt (Stanford University), and Anita Spielman (Williams), seems to have generalized the result to equal-volume double bubbles in four-dimensional space for example.

EPISODE 24

Shortest Road Networks

Question 1. What is the shortest road network connecting the four corners of a unit square?

Answer. The "double Y" of Figure 39 with length $\sqrt{3} + 1$, or about 2.73, beats the popular answer of the X of length $2\sqrt{2}$, or about 2.83, as well as the U of length 3. It is a general principle that at junctions in shortest networks the roads always meet in threes at angles of exactly 120 degrees, just as soap bubbles do, for the similar reason of minimizing surface energy. Perhaps our interstate highways should meet in threes to save concrete, and decrease congestion as a side benefit!

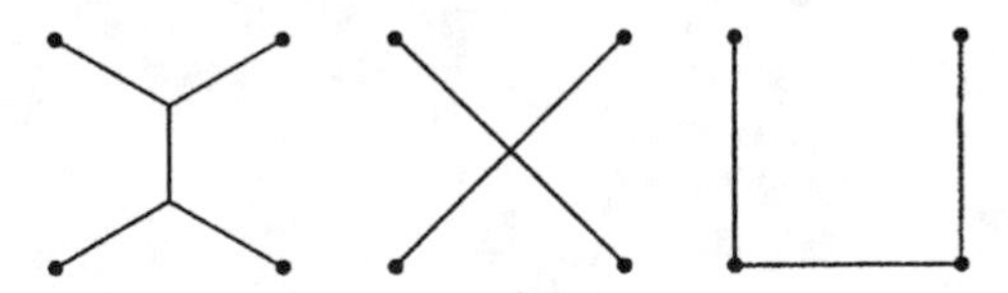

FIGURE 39 The "double Y" is shorter than the X for connecting the four corners of a square.

Question 2. What is the shortest road network connecting the six vertices of a regular hexagon?

Answer. The shortest road just traces five of the sides of Figure 40, and can get you from any vertex to any other, although by a very circuitous route from the upper left to the upper right. The alternative pictured branching network has length 6.

Jarnik and Kössler proved this result in 1934 for the regular hexagon, and for all regular n-gons with $n \geq 13$. It was another fifty years until Du, Hwang, and Weng proved the result for all $n \geq 6$.

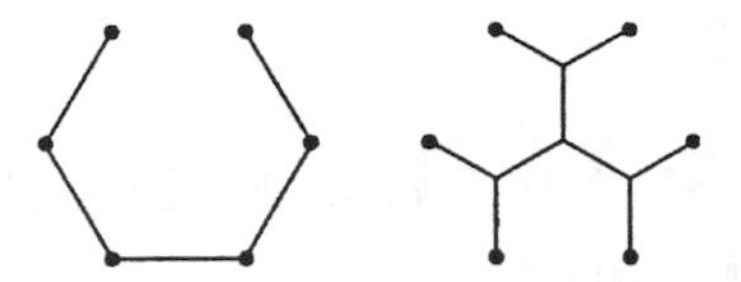

FIGURE 40 The shortest way to connect the vertices of a regular hexagon is to use five of the six sides.

When this appeared in the "Math Chat" newspaper column, I concluded, "If you can find a simple proof, you win instant mathematical fame."

Sure enough, a year later, reader and mathematician Gary Lawlor of Brigham Young University announced a wonderful new simple proof for the hexagon at the Lehigh University Geometry/Topology Conference. He is interested in pursuing his ideas with students and others. His email address is lawlor@math.byu.edu.

EPISODE 25

Can Three States Meet at More Than One Point?

When I was a student at MIT, my roommate's brother, a Mathematics Ph.D. student at MIT at the time, used to drive us around and entertain us with all kinds of questions and stories. Here is one of my favorites:

Question. (Robert Kimble). Can you name three U.S. states that all meet at three different points?

My own Williams College lies in Williamstown, in the extreme northwest corner of Massachusetts, where Massachusetts, New York, and Vermont all meet at a point. Maryland, Virginia, and the District of Columbia all meet at two points. But could three states possibly meet at three points?

Answer. Mike Bevan proves that this cannot happen unless one of the states is disconnected. Mike Tupper writes, "I figured the Mississippi River had to be involved in such a ridiculous thing occurring, and sure enough, that's where I found it. The [north-south] border between Tennessee and Kentucky [follows] a straight line," with Missouri to the west on the other side of the river. James Fahs continues, "The Mississippi takes a sharp turn, called the New Madrid Bend," crossing the Tennessee–Kentucky border line three times. Tupper continues, "A little piece of Kentucky gets completely separated from the rest, causing there to be three distinct points shared by these three states." (See Figure 41.)

John Connolly reports that the river's path was changed by the "New Madrid earthquakes," 1811–1812, among the most violent ever to hit the United States. According to a description of *On Shaky Ground* by Norma Hayes Bagnall on the University of Missouri Press homepage (www.system.missouri.edu/upress), "Vibrations were felt from the

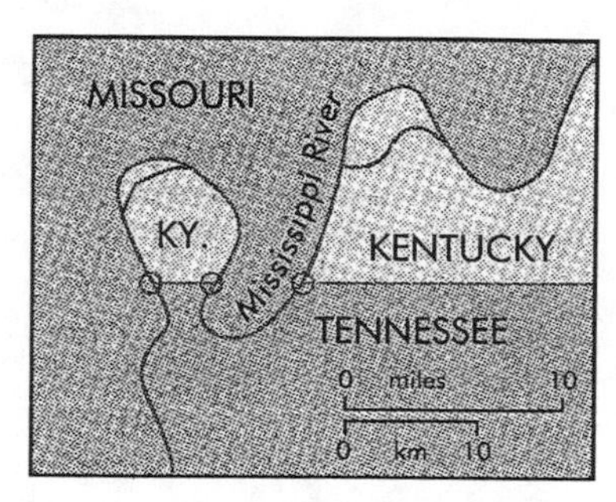

FIGURE 41 A little disconnected piece of western Kentucky makes it possible for Kentucky, Tennessee, and Missouri to all meet at three different points (*The Christian Science Monitor,* May 14, 1998.)

Rocky Mountains to the Atlantic coast and from Mexico to Canada.... Five towns in three states disappeared, [new lakes appeared], and the river flowed backward for a brief period."

Incidentally, it is interesting to look for countries that are not connected, such as the United States (Alaska and Hawaii), Russia (Kaliningrad), and Denmark (Greenland). There are even some countries that are connected but not simply connected (have holes in them), such as Italy (Vatican City is one hole) and South Africa.

Fifth Story: Physics and the World

EPISODE *26*

Balls in the Air and Falling Elevators

Somehow we had lots of fun in my high school physics class, with my good friend Randy Kimble (who ended up becoming a rocket scientist) and the class genius Peter Friedland, even if our teacher, Mr. Hayes, was a bit past his prime. At one memorable period, Mr. Hayes discovered a new kind of instructional text, designed for him to read sentences and for the class to complete them. The development was so pedantic and predictable that we amused ourselves by finishing instead the next sentence that he had not even read yet. Mr. Hayes would frown and shake his head, tell us the correct answer, go on to the next sentence, and then find at the end to his utter bafflement the answer we had already given.

I guess every high school physics class discusses why a ball thrown in the air takes the same amount of time to come down as to go up, neglecting air friction. Somehow we fixed upon that phrase, "neglecting air friction," ritually repeated throughout physics books, and asked what really happens when you take friction into account. It seems confusing at first. Coming down, the ball, slowed by friction, has more time later to gather speed! Peter Friedland consulted a physics professor at the local Muhlenberg College, who provided him with the appropriate differential equation for analysis. But you do not really need differential calculus to answer the question. Peter got the wrong answer anyway. Logic is enough.

The answer is that with friction the ball rises faster than it falls. You have to throw it harder and faster to get it to reach the same height, whereas it comes down slower. Matthew Tarpy explains that on the way up friction "is working in tandem with gravity so it will slow the ball faster." On the way down, friction "will be working against gravity, hence slowing the return." Chuck Gahr points out that friction is always dissipating energy, so that later speeds are slower. John Vance reports that beach balls actually do go up much faster than they come down.

Lawson Harris asks about the effect of friction on the total time in the air. With friction, although you do have to throw the ball up harder and faster, the delay coming down is a bigger effect, as is easiest to see if the friction is very great, in which case it could take almost forever to come down.

The other objects that often fall from great heights are cartoon characters. Wile E. Coyote is always finding the cliff giving way underneath him. Lynne Lawson wonders why he doesn't save himself by jumping off just before the rock hits the ground. You might object that it would be hard for him to time his jump perfectly, but if he just keeps jumping up and down he'll probably be in the air when the rock hits bottom. To keep things tidy for the column, I put him in a falling elevator (Figure 42).

FIGURE 42 Lynne Lawson wonders if you can save yourself by jumping off a falling elevator just before it hits the ground.

My competitor Marilyn Vos Savant, in her "Ask Marilyn" column (*Parade*, January 18, 1998), said that it is impossible to jump in an elevator in free fall. Actually, it is quite easy, since Wile E. would feel weightless, just as astronauts feel weightless in a freely coasting orbiting spaceship. Pushing up onto his tip toes, he would find himself gently floating up toward the ceiling of the elevator. Pushing off the ceiling, he would float gently back to the floor. From the start it would be hard to jump forcefully, since if he bent his knees, his feet would lift off the floor!

Unfortunately, jumping would do Wile E. Coyote little good. He would still be plummeting downward, just at a slightly slower speed than the elevator. (Perhaps this is what Marilyn meant.) His landing would be just a fraction of a second later, and almost as hard.

Actually, my first draft of the column began like this:

> Suppose, God forbid, you found yourself in a broken, brakeless elevator plummeting downward...

But my editor thought it was too gruesome.

EPISODE 27

Do Airplanes Get Lighter as Passengers Eat Lunch?

Karen Reinecke asks whether an airplane gets lighter when the passengers eat lunch (Figure 43).

FIGURE 43 Karen Reinke asks whether an airplane gets lighter when the passengers eat lunch.

Jan Smit explains that the airplane does not get lighter during lunch as long as everything (including the carbon dioxide and heat produced) stays on the plane. Ruth Gatto rightly attributes weight loss to "any loss of liquid, solid, or gas such as fuel, oil, oxygen, external ice dropping off. . . ." Chuck Gahr and Brennie Morgan mention the weakening of gravity as the plane rises farther from the earth (not to mention the gravitational effect of other heavenly bodies). Also, as the plane accelerates downward you feel lighter, and perhaps you are (depending on how you define weight).

Meanwhile the orbiting astronauts in the space station appear weightless, and therein lies another puzzle. After all, the space station is only about 6 percent farther from the center of the earth than the earth's surface. Therefore gravity should be just about 12 percent less than on the earth's surface (by Newton's "inverse square" law of gravity, if you like). Then why will the astronauts appear weightless?

The answer is that the orbiting astronauts appear weightless because the whole station is in free fall. If the station continued straight ahead, it would shoot off into space, but fortunately it is always falling back toward earth at precisely the right rate to balance that effect. As Jason Heine puts it, "An astronaut floating in the middle of a room will be falling but the room will also be falling at the same rate."

W. Cullers gives an alternative explanation in terms of the apparent "centrifugal force" pulling circling objects outward, which you feel when your car speeds around a corner. For the orbiting astronauts, centrifugal force balances gravity, and they appear weightless.

Ed Dravekcy reports that the movie *Apollo 13* filmed weightless scenes in one of NASA's special KC-135 planes. When the plane is zooming upward at a 45-degree angle, power is turned off for about 15 seconds of free fall. The plane coasts upward, levels off, and falls downward in a "parabolic" path. According to an old web page, "overall, cast and flight crew flew 612 parabolas for a total of 3 hours and 54 minutes of weightlessness."

Mike Bevan points out that "you will have a very small apparent weight in the space station": astronauts in the part of the station closest to the earth will have a slight pull toward it. "This is a very small effect, though important to [certain delicate] experiments." Likewise, while the earth is in orbiting free fall around the sun, the parts closest to the sun feel a small tug, and this contributes to the tides.

We conclude with a question about weight that created more vehement disbelief among newspaper column readers than any other.

Milk Bottle Challenge (Michael Marcotty). Take an old-fashioned bottle of nonhomogenized milk, with the cream risen to occupy the narrow neck at the top, shake it to mix in the cream, and put it back down on the table (Figure 44). Is the pressure of the milk on the bottom of the bottle the same as before mixing, or greater, or less?

The obvious answer is that the pressure on the bottom must remain the same, because the weight is just the pressure times the area, and the weight of the milk does not change. The obvious answer is wrong!

Greg Chapman rightly reports that the pressure on the bottom will increase. The pressure at the bottom depends on the density of the fluid at all levels. Since a given amount of fluid in the narrow neck covers

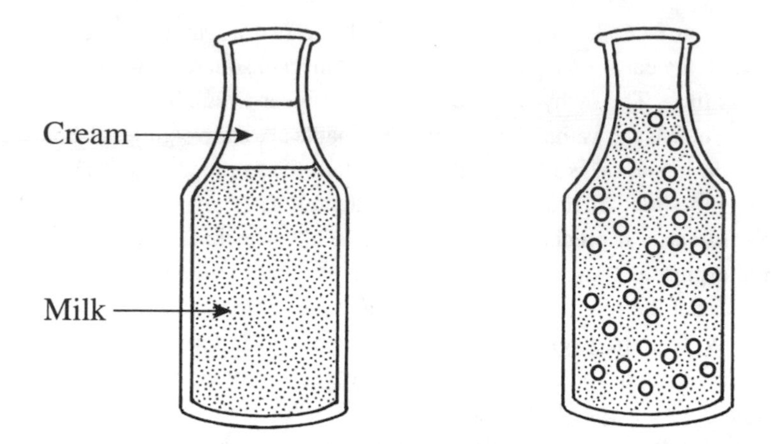

FIGURE 44 Can shaking up a milk bottle change the pressure on the bottom?

more levels, it counts more. When the lighter cream occupied the neck, it counted more, and the pressure was less.

A storm of protest followed the appearance of this (correct) answer in the newspaper column:

> "Your solution to the Milk Bottle problem is a real howler."
>
> "If shaking the milk doesn't change its weight, how can the pressure on the (flat, horizontal) bottom of the bottle change?"
>
> "Congratulations on inventing an anti-gravity machine!"

So what is wrong with the analysis by weight? It is of course true that the weight does not change. But the net weight is the downward pressure on the bottom *minus the upward pressure on the sloping sides below the neck.* When you shake the bottle, the total pressure inside increases, there is more pressure downward at the bottom, and exactly the same amount of extra upward pressure near the top.

Professor Vernon C. Matzen is making this problem a quiz for his mechanics course at North Carolina State University.

EPISODE 28

Tides and Spinning Sprinklers

Water Level Question 1 (thanks to Jeff Bradford). When you throw the anchor over the side of your boat, does the water level in the pond rise or fall?

Answer. This is a hard question because lightening the boat lowers the water level, but adding the anchor to the water raises the water level.

Imagine someone on the dock first removing the anchor from the boat and then throwing it in the water. When she lifts it, the boat rises and the water level falls an amount depending on the anchor's weight. When she throws it in the water, the level rises an amount depending on the anchor's volume. Since anchors weigh more than the same volume of water (that's why they sink), the effect of removing the anchor from the boat is greater, and there is a net fall in water level.

Other helpful readers ask what happens if the anchor floats! Or does not reach the bottom!! (No change in water level.)

Question. When ice melts in a glass of warm water, does the water level rise or fall? (Figure 45.)

Although ice is less dense than water (that is why it floats), the volume that matters is the volume after it melts. Since the volume of melted ice is of course identical with the volume of water (of the same weight), the water level in the glass stays about the same.

Curiously enough, although water expands when it freezes, it contracts when cooled (down to couple of degrees above freezing). Since the melting ice cools the water, the water contracts and the level actually falls slightly. Jan Smit concludes that "the anticipated rise of the ocean

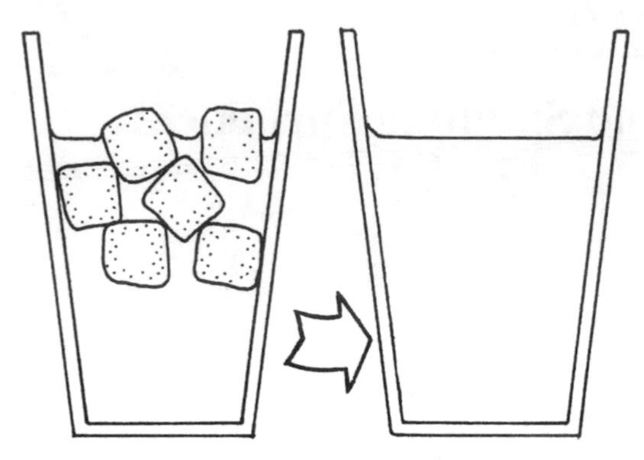

FIGURE 45 When ice melts in a glass of water, does the water level rise or fall?

level due to global warming can therefore be only due to the ice resting on the bottom."

That little comment by Smit in the newspaper column caused quite a stir. The melting ice "resting on the bottom" must include ice on land as well as huge chunks extending all the way to the bottom (and certainly not smaller chunks of ice sunken to the bottom—ice floats!). Also, ice flows off the land into the oceans. There is another comparable cause of rising oceans: the ocean water expands as it gets warmer.

Question. (thanks to John Kruschke). Which makes the water level in a bucket rise more, adding a pound of salt or a pound of sand?

Answer. The sand, because the salt dissolves into the water and occupies less space.

Mount Everest. Although Mt. Everest, at 29,028 feet above sea level, is the highest mountain in the world, it is not the farthest from the center of the earth. The earth's bulge at the equator pushes Chimborazo in Ecuador, at 20,561 feet above sea level, farther. (How much?) Now suppose you run a water pipe from Everest to Chimborazo. Which way would the water flow? (Figure 46.)

Answer. Although it seems hard at first to determine water level equilibrium on a bulging earth, the ocean does that for us! Since Everest is higher above sea level, the water would flow from Everest to Chimborazo. Jim Henry remarks that such a pipe along the ground would be over

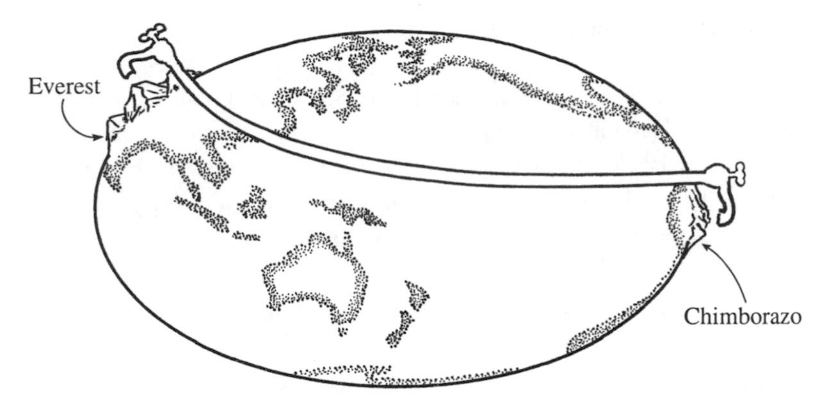

FIGURE 46 The earth's bulge at the equator pushes Chimborazo in Ecuador farther from the center of the earth than Mt. Everest. If you ran a water pipe between them, which way would the water flow?

10,000 miles long and that the pressure at sea level would be over 800 times atmospheric pressure. Incidentally, Chimborazo is about 7,000 feet farther from the center of the earth than Everest.

Thomas Seager gave his students this question on an exam. A typical student answered incorrectly that, "Gravitational force ... would be stronger at Mt. Everest because it is closer to the center of the earth. The water would flow toward the stronger pull of gravity, [toward] Mt. Everest." Seager reports that "most students felt water would flow toward Everest, but that the atmosphere would be thinner there. These are inconsistent conditions, and I have a little work to do in class to explain why.... Correct answers were deduced from the fact that sea level must represent a surface of equal potential, and that this is the correct reference for determining potential energy of water in the pipes on each mountain peak."

Incidentally, Mt. Olympus on Mars is about 90,000 feet high.

Tides. Probably the most noticeable changes in water levels are the tides. Gregory Sahagen notes that the sun's gravitational force on the earth is greater than the moon's and asks why the tides are primarily due to the moon.

Answer. The answer is that tides are due not to the magnitude of gravitational force but on the variability of gravitational force, pulling harder on nearer waters (and causing them to swell), average on middle waters (which remain relatively low), and weaker on distant waters (and letting them swell in the opposite direction). Since the moon is closer than the sun, the percentage change in distance from one side of the earth to the other is greater, and as it happens this effect is enough to make the moon's tidal effect greater. Note that since the tides bulge in both directions, there are two high tides a day, not just one.

We close with the famous question of Feynman's sprinkler.

Feynman's Sprinkler. An old-fashioned water sprinkler with S-shaped spiral arms is propelled counterclockwise as the water spurts out. Which way will it turn if you place it under water and have it suck water in? (Figure 47.)

Sirius Fuller answers that the sprinkler "will not turn at all. It may vibrate, but it won't spin. The reason I know this is that I read a book entitled *Genius* by James Gleick. It's about the life of Richard Feynman.

FIGURE 47 Feynman wondered which way an underwater sprinkler sucking in water would turn.

"At one part it tells how Feynman, and several others, were trying to solve [this very] problem.... After much debate Feynman went to a lab and put the sprinkler in a sealed glass tank mostly filled with water. The sprinkler had a tube attached to it going out of the tank, through the seal, and into a different container. He then forced air into the top of the tank so that the water would be forced down into the sprinkler and up the tube. The sprinkler started to vibrate. He increased the amount of air going into the tank, yet the sprinkler did not spin. Feynman continued to increase the air pressure until the glass tank shattered. Besides getting him permanently banned from the lab, it also proved what he initially suspected, which was that the sprinkler would not move. After thinking about it he put forth the hypothesis that water was going into the sprinkler from many different angles. The different forces canceled each other out, and thus the sprinkler would not spin."

Although most readers agreed with this published answer, Erik Randolph, for example, argues that conservation of momentum requires that if the water in the sprinkler is going one way, the sprinkler must turn the other way. Leaf Turner points out that this backwards effect, much smaller than that due to all the water spurting out in normal sprinkler operation, and almost undetectable, has been measured in experiments with a special, nearly frictionless sprinkler. See Berg and Collier, *American Journal of Physics,* Vol. 57 (1989), p. 654, and references to previous articles with all possible different conclusions about what would happen and about what Feynman thought would happen.

It is too late to ask Feynman, but the controversial accounts of Feynman's sprinkler problems also involved his mentor and friend Prof. John Wheeler, who appeared at my lunch table at Prospect House at Princeton University on December 16, 1997. Of course I asked him about the experiment and what Feynman thought. What happens? He said, "After an initial jolt, nothing."

EPISODE 29

Cars and the Future

I still remember my first 70-miles-per-hour speed limit sign. I felt like I was driving my mom's Camaro into the future. Then came the oil crisis, and 55-miles-per-hour limits. Marc Abel asks, When a state increases a highway speed limit from 55 to 65 miles per hour, by what percent does the road capacity (in cars per hour) change?

Jean Hunter explains that if traffic moved at the same spacing at both speeds, then the capacity would increase by the same amount, $\frac{10}{55}$, or about 18 percent. However, if the traffic spacing is governed by the safety rule of one car-length between cars for every 10 miles per hour of speed, then capacity increases only about 2.5 percent.

According to Glenn Grigg, a traffic engineer (which he finds "an interesting and rewarding" job), "the Highway Capacity Manual states that the ideal capacities of 2,200 and 2,300 vehicles per lane per hour can be sustained over a broad range of speeds." Scott MacCalden reports that "for freeway driving during commute periods (when almost all vehicles are passenger cars, and the drivers are familiar with the road) average time intervals or 'headways' between cars may vary between 2.0 and 1.5 seconds. This is much more than on the Indy 500 track where headways can be considerably less than 1 second and average speeds reach 200 mph!"

Jan Smit observes that at high speeds, where safe braking distances become more important than reaction time, road capacity might decrease as speeds increase.

Fuel efficiency will doubtless become increasingly important.

Question. You are out driving your gas-powered or perhaps new electric battery-powered car. At a freeway entrance, you need to accelerate from 0 to 60 miles per hour within 20 seconds. What do you think is the best way to do this to use the least amount of energy?

Charles Sullivan reports that engines are more efficient at full throttle at high gears, so at least with a stick shift you should give it plenty of gas but shift up as quickly as possible. Marc Abel reports that batteries are more efficient at low power, so that it is better to apply more thrust at lower speeds. John Morrison duly notes the first, most obvious rule: don't step on the brake!

Question. Starting from scratch, what would be the best transportation system of the future for getting everybody to work?

Mark Thompson describes trains that do not waste any time stopping at stations. The last little car on the train separates itself, stops, exchanges passengers, and then accelerates to be joined to the front of the next train through the station.

Eric Klieber has an idea for luring car drivers back to public transportation: "There's only one thing Americans love as much as cars, and that's TV. [Put] TVs in every bus and rapid transit car; TVs in every waiting area; even TVs moving in sync with escalators and moving sidewalks." Everyone could wear radio headphones. TV networks could pay for the whole thing. Actually, we think one could do better than network TV; perhaps some "Math Chat" feature!

Adam Snow may have the best transportation system of the future—SLIPPERS: "Since the ideal job of the future would allow one to telecommute, the ideal transit system from bedroom to computer terminal would be a comfortable pair of slippers." (Figure 48.)

John Robertson, looking farther ahead, proposes instant teleportation (as on "Star Trek"), and having robots do the work, so we can spend our time doing mathematics.

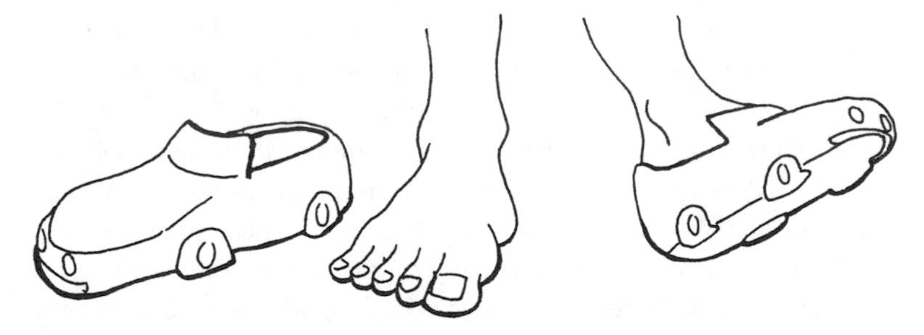

FIGURE 48 The ideal transportation system of the future may be—a comfortable pair of slippers.

EPISODE *30*

Eclipses and Mirrors

Eclipses of the sun and moon are relatively rare because the moon's orbit is tilted, carrying it out of the plane of the earth's orbit. Otherwise, we would have a lunar eclipse at every full moon and a solar eclipse at every new moon. Aubrey Dunne rightly points out that worldwide there are more solar eclipses than lunar eclipses. Then why do lunar eclipses seem so much more common?

Answer. A lunar eclipse is visible over more than half the earth, while a solar eclipse is visible only in a relatively narrow band. Hence for any particular spot on earth, lunar eclipses are much more common.

A woman called our local radio station WKAP in Allentown, Pennsylvania, concerned about the warnings about damaging sight by staring at the sun during an eclipse. She began, "I have a question about the eclipse. If this is so dangerous, why are they having it?"

Mirror Challenge (Fred Wedemeier). Why do mirrors reverse left and right and not up and down? (Figure 49.)

This is one of the most confusing questions I know, and there are some very ingenious wrong answers: "The mirror does reflect from top to bottom, but your brain recognizes that the shape should be right-side-up, and flips it for you." "The image in a mirror is rotated about the vertical axis because a person, when turning to regard another person, rotates himself about the vertical axis." (But if we stood on our heads to look behind us, would mirrors suddenly start working differently?) "This is just the way your eyes see it. If your eyes were one above the other, the mirror would reflect up and down." (But if you close one eye, do mirrors work differently?)

FIGURE 49 Why do mirrors reverse left and right and not up and down?

Bill Cutler rightly explains that mirrors "don't reverse left and right; they reverse front and back. Stand facing north with a mirror in front of you. Wave your left (west) hand. The image in the mirror waves its west hand too, so there is no left-right reversal. However, your nose is pointing north while the nose of the image is pointing south." This is what makes you interpret your reflection's west hand as a right hand. See the excellent discussion in Martin Gardner's *The New Ambidextrous Universe* (Chapters 1 and 3, especially page 19 in the third revised edition, 1990).

Math Chat Winners

"Math Chat" Winners as of December 1, 1999

Abel, Marc, 100, 101
Amaral, Joseph
Anderson, Kurt
Annekov, Vladimir
Arrow, Holly
Baars, Luis, 70
Baisch, Joseph
Barbalace, Dan, 16
Barnett, Brian
Beck, Jozsef
Bergstrand, Betty
Bergstrand, Deb, 70
Berkovitz, Joe
Berman, Noah
Berquist, Denis
Bevan, Mike, 13, 84, 93
Bird, Alex
Bitzenhofer, Neil
Bliss, Robert, 57
Bond, William, 42
Bourgoin, Mario
Bradburd, Rebecca
Bradford, Jeffrey P., 94
Brahinsky, Eric, 48, 57, 58
Brakke, Ken
Brandt, John
Bredt, Jim, iii, 57
Bright, Ted
Brill, Stephen
Buchanan, Greg
Bunch, Charles
Burkhart, Reed, 42
Burns, Christopher, 34
Cañizo Rincón, José
Carlson, Joan
Carmichael, Jean-Pierre
Case, Glenn
Catagnoli, Neal
Catapano-Friedman, Rebecca
Cebada, Daniel, 42
Chace, Charles
Chandler, Matt
Chang, Thomas
Chapman, Greg, 93
Charters, Duncan
Chkhenkeli, Misha, 19
Clark, Dennis
Clark, Tim
Cobi, Bruce

Index